Conservation in Common

GEOGRAPHIES OF JUSTICE AND SOCIAL TRANSFORMATION

SERIES EDITORS

Mathew Coleman, *Ohio State University*
Ishan Ashutosh, *Indiana University Bloomington*

FOUNDING EDITOR

Nik Heynen, *University of Georgia*

ADVISORY BOARD

Deborah Cowen, *University of Toronto*
Zeynep Gambetti, *Bogaziçi University*
Geoff Mann, *Simon Fraser University*
James McCarthy, *Clark University*
Beverley Mullings, *Queen's University*
Harvey Neo, *Singapore University of Technology and Design*
Geraldine Pratt, *University of British Columbia*
Ananya Roy, *University of California, Los Angeles*
Michael Watts, *University of California, Berkeley*
Ruth Wilson Gilmore, *CUNY Graduate Center*
Jamie Winders, *Syracuse University*
Melissa W. Wright, *Pennsylvania State University*
Brenda S. A. Yeoh, *National University of Singapore*

Conservation in Common

MANAGING WILDLIFE AND SUSTAINING
COMMUNITY ON THE MAASAI STEPPE

JUSTIN RAYCRAFT

THE UNIVERSITY OF GEORGIA PRESS
Athens

© 2025 by the University of Georgia Press
Athens, Georgia 30602
www.ugapress.org
All rights reserved
Set in Minion by Westchester Publishing Services

Most University of Georgia Press titles are available from popular e-book vendors.

Printed digitally

EU Authorized Representative Easy Access System Europe—Mustamäe tee 50, 10621 Tallinn, Estonia, gpsr.requests@easproject.com

Library of Congress Cataloging-in-Publication Data

Names: Raycraft, Justin, 1990– author
Title: Conservation in common : managing wildlife and sustaining community on the Maasai steppe / Justin Raycraft.
Other titles: Geographies of justice and social transformation 71.
Description: Athens : The University of Georgia Press, 2025. | Series: Geographies of justice and social transformation ; 71 | Includes bibliographical references and index.
Identifiers: LCCN 2025020543 | ISBN 9780820374789 hardback | ISBN 9780820374796 paperback | ISBN 9780820374802 epub | ISBN 9780820374819 pdf
Subjects: LCSH: Wildlife conservation—Social aspects—Tanzania | Wildlife management areas—Social aspects—Tanzania | Community-based conservation—Tanzania | Human-animal relationships—Tanzania
Classification: LCC QL84.6.T34 R39 2025 | DDC 333.95/41609678—dc23/eng/20250709
LC record available at https://lccn.loc.gov/2025020543

In memory of my grandmother, Polly Baria (1932–2024), whose stories about her life in Tanzania inspired me to become an anthropologist and find my way back there.

CONTENTS

List of Illustrations   ix

List of Abbreviations   xi

Introduction   1

CHAPTER 1   Maasai Society and the State   24

CHAPTER 2   The Lolkisale Land Squeeze   44

CHAPTER 3   Politics of Hunting and Photographic Tourism   64

CHAPTER 4   Creating a Wildlife Management Area   81

CHAPTER 5   The Rise of Randilen   103

CHAPTER 6   Foundations of a Social Enterprise   126

CHAPTER 7   Complexities of Community-Based Conservation   151

Conclusion   173

Acknowledgements   185

References   189

Index   203

ILLUSTRATIONS

Figure

Community attitudes toward Randilen WMA   105

Map

Location of Randilen Wildlife Management Area (WMA) in northern Tanzania   21

Photographs

A Kisongo Maasai woman smiling in Lemooti village   17
An Arusha man in Lolkisale village with his herd of goats and sheep   18
A Kisongo Maasai man guiding the anthropologist to key sites in Lemooti village   45
Farmlands enclosing Lolkisale village   45
Young men pushing a vehicle out of the mud on the road to Lolkisale   60
An elephant in Randilen WMA   61
A giraffe in Randilen WMA   62
Wildebeest near Sunset Hill in Randilen WMA   63
Local cattle under the Randilen WMA highway sign near Naitolia village   108
Randilen WMA authorized association members meeting in Makuyuni village   119

List of Illustrations

An Arusha woman in Oldonyo village threshing beans with a stick   127
Honeyguide and Randilen WMA staff helping a woman in
    Nafco village   131

Table

Community attitudes toward Randilen WMA disaggregated by member
    village   106

ABBREVIATIONS

AA	Authorized Association
AWF	African Wildlife Foundation
CAMPFIRE	Communal Areas Management Programme for Indigenous Resources
CBO	community-based organization
CCROs	Certificates of Customary Rights of Occupancy
DGO	District Game Officer
EASTCO	East African Safari & Touring Company
GCA	Game Controlled Area
GEF	Global Environment Facility
HEC	human-elephant conflict
LCA	Lolkisale Conservation Area
LLWZ	Lolkisale Livestock and Wildlife Zone
LVC	Lolkisale Village Council
MDC	Monduli District Council
MEDA	Makuyuni Elephant Dispersal Area
MNRT	Ministry of Natural Resources and Tourism
MODECO	Monduli Development Corporation
MP	Member of Parliament
MSP	Multiple Stakeholder Partnership
NAFCO	National Agricultural and Food Corporation
NCA	Ngorongoro Conservation Area
NCAA	Ngorongoro Conservation Area Authority
NGO	nongovernmental organization
NP	National Park
OBC	Ortello Business Corporation
RZMP	resource zoning management plan

TANAPA	Tanzania National Parks Authority
Tarangire NP	Tarangire National Park
TAWA	Tanzania Wildlife Management Authority
TCCL	Tarangire Conservation Co. Ltd.
TPMZ	Tourism and Photographic Management Zone
TRA	Tanzania Revenue Authority
UCRT	Ujamaa Community Resource Team
VGS	village game scouts
WCA	Wildlife Conservation Act
WMA	Wildlife Management Area

Conservation in Common

Introduction
Conservation in Context

This book is about conservation—a set of ideas and practices meant to protect Earth's natural resources for posterity by reducing human impacts on the environment. Efforts to manage anthropogenic pressures on ecosystems and promote sustainable resource use are neither new nor unique to the Western intellectual canon. Indigenous peoples have long embraced values of respect and reciprocity in their forms of relationality with the environment, establishing over millennia carefully tuned customary institutions for regulating access to common pool resources. Conservationism has a much shallower history. Seeded by Enlightenment era thinking, conservation emerged as a policy paradigm in the late nineteenth century in response to mounting exploitation of the planet's natural resources due to the rise of global capitalism in centuries prior. With the growth of capitalism came accelerating colonial expansion and a new way of thinking about the natural world in terms of commodities that could be manipulated to fit the objectives of a Eurocentric worldview of modernity (Wolf 1982). Notions of progress drove technological innovation and scientific advancements, in turn reshaping the productive capabilities of human society and driving industrialization. To accommodate rapid increases in production in western Europe, colonial powers took advantage of mercantile links with frontier nations rich in raw materials and untapped labor forces, creating templates of structural inequality that still endure today. Unlike Indigenous worldviews, the Western conservation movement is born of this same history of capitalism and colonialism that produced environmental disrepair in the first place (Büscher et al. 2012; Fletcher 2023). Somewhat paradoxically, it attempts to address the environmental consequences of development and extractivism without taking stock of their root cause—a globalized political-economic system from which conservation itself emerged (Büscher and Davidov 2013; Büscher and Fletcher 2015).

The field of conservation has been undergoing growing pains since its origins, at times with little reflexive consideration of its own messy place in a larger political and economic apparatus that has become entrenched at the global level (Vaccaro et al. 2013). Critical political ecologists offer crucial insights into these entanglements by bringing attention to historical distributions of power (Bluwstein 2022). While conservation has the potential to address patterns of environmental exploitation wrought by economic growth and political progress, for instance, it is also capable of reinforcing social inequalities by dispossessing local and Indigenous communities who care deeply about environmental sustainability (Brockington and Igoe 2006). Since the formation of Yellowstone National Park in 1872, national parks have become synonymous with conservation, tourism, and recreation (Igoe 2004). Underpinning the imperative to form parks and protected areas restricting human activities has been the social construction of nature and society as a binary opposition that can be tidily sundered through boundary-making (Cronon 1996; Nash 2014). The nature-culture dualism structuring conservationist thought—not unlike Descartes's attempts to parse mind and body—has failed to grasp the simple fact that people have never been outside nature and never will be (Descola 2013; Williams 1980). Indigenous peoples who have historically inhabited the planet's most biodiverse landscapes—many of which have now become classified as world-renowned protected areas—have not lost sight of this basic premise (Goldman 2020; Mulrennan et al. 2019).

Despite its conceptual flaws, the protected area model of conservation has expanded around the world with remarkable efficiency, proliferated largely by global conservation organizations, ecotourism sector investors, and governments (Brockington et al. 2012). Across sub-Saharan Africa, the rise of protected areas as the dominant model of conservation since colonial contact has been profound. These trends are especially pronounced in Tanzania, where more than 40 percent of the country's roughly 886,000 km² landscape has been designated as part of a protected area, a particularly significant portion considering the country's population of about 67 million people. Ironically, protected areas emerged in Tanzania around the turn of the twentieth century mainly as a conservation "solution" to the settler problem of unsustainable recreational safari hunting, which followed centuries of commercial ivory exploitation by Arab caravan traders (Nelson et al. 2007). Put differently, protected areas became a way to render technical a problem that was rooted in the political economy of colonialism.

Social scientists studying conservation in Tanzania have been especially mindful of this troubling history by attending to power dynamics underlying the formation of protected areas and the uneven distributions of conservation costs and benefits (Brehony et al. 2022; Homewood et al. 2012, 2020; Kamat 2024; McCabe and Woodhouse 2022). In postcolonial contexts like Tanzania, overly simplistic narratives about the uniform benefits of conservation can translate into the top-down establishment of protected areas that exclude and marginalize local communities both geographically and institutionally (Bluwstein and Lund 2018). Environmental crisis narratives pushed by the state and conservationists add further justification for displacing people while masking complex sociopolitical histories and patterns of local resource use that were never antithetical to the aims of conservation to begin with (Benjaminsen and Bryceson 2012). Ideologically informed by the idea of a wild nature "out there" in need of preservation from the threats posed by human society, conservation in Tanzania has a long history of evicting local communities from wildlife-rich areas, a trend that is unfortunately still extant (Brockington 1999; Weldemichel 2020, 2022). Conservation has thus become a self-fulfilling prophecy in Tanzania that reifies through political intervention the very thing it seeks to protect: an unpeopled wilderness (Neumann 1998). Brockington (2002) refers to this erasure of human histories of presence and stewardship through the formation of protected areas as "fortress conservation." Unsurprisingly, in a contemporary context, those calling for local communities to be pushed off their traditional lands—namely, the state and foreign investors—have stood to benefit from fortress conservation through monopolization of substantial safari tourism revenue collection (Gardner 2012; Igoe 2017).

Conservation scientists have historically been complicit in these processes by focusing their gazes exclusively on wildlife or grassland dynamics without adequate recognition of the dynamic role that people have played throughout history in shaping ecosystems (Mbaria and Ogada 2016). The far-stretching savannas of the Serengeti, for instance, which have become iconic of the unpeopled nature that conservationists seek to protect, have likely become viewable as they are today through hundreds of thousands of years of human interaction with landscapes by means of fire and, later, livestock grazing (Shetler 2007). Scientists, however, routinely overlook these historical patterns of presence and stewardship, allowing the hegemonic influence of "wilderness" to endure in conservation discourse as a socially constructed place of mind.

Established norms within the field of rangeland science held until about fifty years ago that savannas had fixed natural carrying capacities and would degrade rapidly if pushed beyond these thresholds. This basic equilibrium framework was bolstered by Hardin's (1968) now infamous "tragedy of the commons" theory, which posited that in the absence of formal governance mechanisms, livestock keepers would overstock shared pastures to maximize their own individual benefits since the costs would be diffused across all resource users. The theory has since been debunked or, rather, refined to refer more specifically to instances of "open access" property regimes where there are no institutions in place at all for regulating access to and use of resources (Hardin 1994). Unsurprisingly, groups of people who depend on a particular resource pool for livelihood often implement informal rules specifying rights, appropriate uses, and responsibilities of the resource users (Bromley 1992; Ostrom 1990). Such systems, which are the very fabric of herding societies in East Africa, are referred to in environmental anthropology as common property systems.

The uninterrogated assumption that herders overstock rangelands for personal benefit and push them past their fixed carrying capacities fits tidily with a conservationist nature-society dualism (Homewood and Rodgers 1987). The political implication of this way of thinking is that pastoralists invariably degrade ecosystems and must be removed from biodiverse landscapes altogether (Turner 1993). Classical rangeland sciences thus stabilize the political dynamics of fortress conservation in Tanzania by providing a technical rationale that buttresses ideology.

Range sciences have since broadened through recognition of dynamic interplays between social and ecological systems and the nutrient cycles that human activities like livestock grazing afford (Behnke and Scoones 2003). Highly variable patterns of rainfall in East Africa necessitate ecological systems that exist in multiple forms of stability depending on changing environmental conditions (Homewood 1994). East African rangelands, it would seem, were never equilibrium systems to begin with, but ones that were constantly adapting to external factors like rainfall, fire, and grazing pressure (Reid 2012). Disequilibrium theorists recognize the fact that herding societies do not unequivocally degrade landscapes and often contribute to sustainable rangeland management through customary institutions like seasonal mobility and rotational livestock grazing (Homewood 2008; McCabe 2003a, 2003b). Unfortunately, the prevailing notion that conservation ought to take the form of bounded protected areas that keep out local people and livestock remains a normative way of thinking in Tanzania, "co-produced" by a history of scientific misunderstanding (Jasanoff 2004).

Toward Community-Based Conservation

Recognition of the shortcomings of fortress conservation in Africa since the 1980s has sparked calls for the devolution of conservation authority to better represent the interests of local and Indigenous communities (Jones 2006). The move toward community-based models of conservation has resulted in varied social and ecological outcomes across the continent. Some initiatives, like Zimbabwe's "Communal Areas Management Programme for Indigenous Resources" (CAMPFIRE) and Namibia's decentralized model of communal conservancies, have shown great promise for reconciling wildlife conservation with the protection of community livelihoods (Suich 2010; Welch 2018). In Tanzania, the shift toward community-based conservation coincided with a move away from single-party socialism and a period of structural adjustment that introduced neoliberal economic reforms and ushered in flows of foreign investment (Neumann 1995). By this time, limitations of the protected area model of conservation had become apparent in Tanzania, as large mammals required much more habitat than had been circumscribed by national parks (Bluwstein 2018b). About half of the total elephant home range in Tanzania, for instance, exists outside protected areas, with elephants moving fluidly across unfenced park boundaries, depending on seasonal variations in rainfall (Nelson et al. 2007).

Following structural adjustments, tourism operators took interest in the pastoral lands adjacent to major parks on the northern safari circuit, knowing that they hosted significant wildlife dispersals. In Tanzania, community lands are classified as villages, which constitute administrative units for governing and managing land within their jurisdictions owing to a history of villagization during the socialist period. Villages offered appealing investment opportunities for tour operators given the looser conservation regulations outside national parks. Pastoralists were often supportive of photographic tourism on village land because communities could negotiate the terms of contracts with investors directly, through their own local governance institutions (Gardner 2016). Generally, agreements involved establishing concessions where crop cultivation would be prohibited but livestock grazing would be allowed, except in areas close to camps or lodges. Villages received direct payments to offset the opportunity costs of not using land for crop production and, in exchange, tour operators gained exclusive rights to host photographic safari tourists in the concessions (Nelson et al. 2010). Perhaps surprisingly to some scholars, neoliberalization thus created opportunities for community-based conservation to blossom in Tanzania in ways that were not possible during the colonial and socialist periods (Gardner 2016). Village-investor relations, of course, were not immune to power dynamics and the potential for investors to take advantage of communities (Goldman 2020; Wright 2019). Nonetheless, mutually beneficial partnerships were also empirically evident in several notable cases (Nelson et al. 2010).

While innovative models of community-based conservation emerged organically in the neoliberal era through direct investments on village land, the state was wary of losing control of valuable tourism revenues. In fairness to the government, private contracts were being negotiated by investors and villages on a case-by-case basis without a formal regulatory structure for taxation. Against this political backdrop, Tanzania implemented reforms to the wildlife sector in the late 1990s based on international trends in community-based conservation elsewhere in Africa. On paper, the policy changes seemed to address the pitfalls of fortress conservation by calling for greater inclusion of local communities in the conservation sector. In practice, they introduced a centralized framework for managing tourism investments on village land (Benjaminsen et al. 2013). These developments were exemplified by the Wildlife Policy of 1998, which seemed to afford local communities legal rights to manage wildlife on their land for their own benefit, taking inspiration from CAMPFIRE in Zimbabwe (MNRT 1998). Unlike CAMPFIRE, however, which ascribed legal ownership of wildlife resources to Indigenous communities, the post-socialist Tanzanian state was unwilling to fully devolve authority to the local level. Following legal reform, the state formally retained legal ownership of land and wildlife within its national borders—a legacy of centralized socialist policies (Shivji 1986). From an analytical perspective, the government's efforts to consolidate central control of wildlife tourism seemed to run counter to the prospect of community-based conservation, which requires a meaningful degree of local participation in conservation practice (Goldman 2003). Recentralization also contradicted the economic principles of neoliberalization, particularly the notion of a "rolled back" state (Wright 2017).

Based loosely on the community conservancy model piloted elsewhere in Africa, the Tanzanian government introduced Wildlife Management Areas (WMAs) in 1998 with formal regulations following in 2002. WMAs involve demarcating some portions of village land as dedicated wildlife habitat in accordance with a land zoning scheme determined by member villages and formally gazetted by the Ministry of Natural Resources and Tourism (MNRT). When they were first introduced, WMAs were framed by the Tanzanian government as community-based models of conservation that formalized the rights of communities to participate in wildlife management while simultaneously conserving natural habitat outside national parks. In practice, they functioned to foreclose direct investments on village land and recentralize collection of wildlife-related tourism revenue outside national parks and game reserves. From a community perspective, WMAs seemed to represent yet another top-down model of conservation that was being forced upon localities from above—a far cry from the ethos of community-led conservation. On legitimate grounds, pastoral communities were initially widely opposed to WMAs, which they viewed through the lens of a long history of fortress conservation as government initiatives to undermine pastoral tenure. Adding to their suspicions, international nongovernmental organizations (NGOs) like the African Wildlife Foundation (AWF) became instrumental in helping the government establish WMAs across village lands, at times with little participatory input from the rural communities whose everyday lives were to be affected (Igoe and Croucher 2007). Anti-WMA sentiments were particularly pronounced in Loliondo, where pastoral Maasai communities rejected the prospect of a WMA altogether (Gardner 2016). Rather than a WMA, Maasai pastoralists wanted more localized forms of community-based conservation that kept power in the hands of villages. Herders sought self-determination, autonomy, and perhaps most significantly, secure land tenure in the face of external interests in grabbing pastoral lands for conservation or other purposes. WMAs seemed to be a way for the state and international NGOs to undercut the village institution for allocating land, which pastoralists viewed as an existential threat to their way of life.

Led by grassroots pastoral NGOs like the Ujamaa Community Resource Team, herding communities sought instead to secure village titles that ensured local control over rangelands. Within villages, individuals or groups could obtain Certificates of Customary Rights of Occupancy (CCROs) to signify legal land holdings. Technically, however, formal land ownership still resided with the state, which leased occupancy to individuals and communities pursuant to their customary rights. Villages and CCROs have been vital for asserting pastoral tenure across northern Tanzania but still run into conflict with state interests when they overlap areas considered valuable for wildlife-related tourism. Ongoing tensions in Loliondo Division over the layered boundaries of villages and a central Game Controlled Area exemplify the precarity of relying on the village institution alone. In June 2022, violent confrontations between local Maasai herders and state paramilitary forces erupted when government officials demarcated the boundaries of Pololeti Game Reserve with beacons, resulting in a dozen arrests of Maasai leaders and a dead police officer. Despite a history of well-received photographic tourism ventures in Ololosokwan and other villages in Loliondo, the state opted for a centralized game reserve that pushed herders off the land entirely. The driving force behind the decision was Ortello Business Corporation, a United Arab Emirates–based company with links to the royal families that generated lucrative revenues for the central government by administering the area as a trophy hunting block for the global elite.

The recent conflicts in Loliondo illustrate the limits of the village institution and CCROs in areas that the state values for wildlife tourism, especially areas with potential for trophy hunting. It would seem that despite an initial period of neoliberalization in the late 1980s and 1990s, the Tanzanian state has since reverted to a stronghanded conservation model by presenting communities with a tacit ultimatum: accept the WMA model of community-based conservation or risk having land grabbed back by the state in partnership with private operators through a fortress approach to establishing parks and reserves that cuts communities out of the equation altogether. WMAs, then, represent a challenging notion for pastoralists living in biodiverse areas to wrestle with. On the one hand, they are part of a centrally devised conservation framework with the potential to undercut local governance institutions, but on the other, they are potentially the lesser of two evils when faced with the prospect of evictions and land alienation. These trade-offs are particularly challenging to navigate since the state has forcibly taken direct investments in villages off the table.

Formalization of WMAs as the de jure model of community-based conservation on village land has thus translated into a conundrum with potentially major consequences for pastoral communities: Do they risk establishing WMAs on village land to access the benefits of safari tourism, or do they reject them and give up on the prospects of participating in the conservation sector altogether? Perhaps more significantly than benefits alone, an underlying question for pastoralists is whether WMAs will dispossess them of their land or provide them with formal tools to articulate their tenure claims. This issue surfaced in Wright's (2017, 2019) ethnographic research with the Maasai community in Longido's Sinya village, which initially refused the prospect of a WMA out of fear of dispossession but later decided to join Enduimet WMA as a means of pushing out a disrespectful photographic tour operator and regaining control over the zoning of a notable trophy hunting block. Wright's work makes clear a consideration that has otherwise not featured into the scholarly discourse on WMAs in Tanzania: although WMAs pose potential threats to pastoral tenure, they can also generate *opportunities* for communities to defend their territories through formalization.

Social scientists, in defense of the community interests they represent, have consistently argued that the WMA framework is largely top-down in structure and bars alternative models of community-based conservation that become possible without state intervention (Kicheleri 2018; Kicheleri et al. 2021). Certainly, it is difficult to view WMAs as genuinely devolved initiatives, as they seem to represent an attempt to recentralize control over the wildlife sector during an era of neoliberalization (Green and Adams 2015). Though seemingly responding to the international trend of establishing community conservancies, the Tanzanian state has demonstrated little contextual flexibility or willingness to foster bottom-up conservation initiatives, making WMAs a difficult sell as consummate examples of community-based conservation in action (Bluwstein et al. 2016). With few exceptions, the scholarly discourse on the social impacts of WMAs has been decidedly critical in this regard, even equating WMAs to the national park model of fortress conservation in some instances, despite their statuses as multiple-use areas (Brehony et al. 2018).

Having Conservation in Common

This book takes up the question of how local community members perceive, interpret, and relate to a WMA in northern Tanzania. The book focuses on Randilen WMA, situated along the northeastern border of Tarangire National Park (Tarangire NP) in an area of crucial ecological significance for elephants. It demonstrates through ethnographic analysis how community members feel about the WMA and, perhaps more significantly, *why* they feel the way that they do. Randilen is a particularly significant case study because its history was shaped by conflicts over the creation of the WMA in 2014 that culminated in open protests and a highway blockade (Loveless 2014). Ten years on from those agitations, however, local opposition to the WMA has given way to nearly universal support. As an anthropologist, I was fascinated by this discursive shift in community attitudes that became apparent when I commenced emplaced fieldwork in Randilen's member villages in 2019. I wondered why people had come to appreciate Randilen in light of its rocky history and entrenched social science critiques of WMAs more generally. While it is tempting to essentialize the sentiments of local people in simple terms, the empirical reality on the ground is unsurprisingly complex. I came to realize over the course of my research that my specific discoveries about community attitudes symbolized the observable tip of an iceberg; the majority of forces at play only became apparent through deeper historical analysis. To paint a more representative picture, this book takes a political ecology approach in laying out the WMA's wider historical context, presented in the form of a narrative that unfolds how Randilen came to be.

I establish in this book the reasons why Randilen WMA has managed to garner local support for conservation and argue that Randilen should be considered a socially successful case of community-based conservation. Making this argument positions me in uncharted territory in the social science discourse on conservation in Tanzania, but the stance I take in this book is less about the uniform generalizability of the WMA model of conservation and more about the social particularities of the Randilen case that offer insight for improving conservation practice elsewhere. While proponents of WMAs—like international NGOs and state actors—have marketed them as "win-wins" that simultaneously benefit wildlife and local communities, there is much to be critical about with this framing. Win-win rhetoric is generally used persuasively as a selling point for an external intervention that is thrust upon communities from above with the expectation that it will reward multiple stakeholders. But such propositions are usually grounded in the economic logics and worldviews of those who are pushing the intervention, and thus often reflect ethnocentric projection. "Benefits" that are narrowly construed in the terms of conservationists or economists may mean little to communities, while costs that community members consider fundamental may be unknowingly cast aside.

This book contributes to the growing scholarly movement on "convivial conservation," which seeks to promote socially and ecologically equitable ways to live together in a multispecies world (Büscher and Fletcher 2019). Rather than trumpeting WMAs as "win-wins," the theoretical intervention I make is around the central question of what it means to find common ground between conservation stakeholders with distinct interests. Establishing commonality is largely an exercise in empathy and involves recognizing that the priorities of local communities are usually not the same as those of government officials, investors, international NGOs, and conservationists. Empathy is not just about sympathizing with feelings. It requires a concerted effort to adopt a different point of view, a process that is fundamental to being human and relating with others. For community-based conservation to prosper, it must be firmly grounded in local understandings and values, and conservation practitioners must be willing to steer management in a direction of the community's choosing. I conceptualize this approach as empathetic conservation.

In unfolding the underlying reasons for the rise of local support for Randilen WMA, I discuss several factors that are specific to the case that may be of significance elsewhere, including the dedicated efforts of grassroots community conservation NGO Honeyguide, the formalization and enforcement of a seasonal livestock grazing plan, the establishment of management units for protecting people's crops from raiding elephants, and the generation of tourism revenue for member villages to invest in local infrastructure development projects. I also draw attention to enduring complexities with the WMA model, including the institutional layering of WMAs over villages, the influential role of village-level governance structures, tensions around revenue sharing between villages, the power of domestic elites, and the ever-present role of the central state in shaping the macropolitical landscape that affects the economic viability of WMAs.

The Randilen case demonstrates that WMAs can work for communities, but this does not necessarily mean that they should be implemented uncritically elsewhere. It is clear in this case that residents of Randilen have embraced the WMA, but this relationship took time to unfurl through thoughtful and patient nurturing from Randilen's management team. Inclusive governance practices generated trust and paved the way for a shared trajectory of social transformation. Randilen now follows a path of its own making, shaped by formal legislation from above and animated by the participation of community members from below. The WMA has provided the frame, and the community continues to paint a picture within it that reflects their values and aspirations. WMAs thus seem to exist somewhere in the space between fortress and community-led conservation depending on the particularities of any given case. In the case of Randilen, there is evidently common ground between the concerns of conservationists who want to protect wildlife habitat, the state, which considers wildlife tourism a key source of revenue for central coffers, and community members whose main priorities are secure land tenure and local agropastoral livelihoods. As exhibited in the title of this book, I propose the phrase having *conservation in common* to describe settings like Randilen WMA, where a conservation territory simultaneously serves different groups of actors with diverse goals, each with stake in determining access to and control of resources and the allocation of duties and rights accompanying their use. As evidenced by the Randilen case, cooperation is likely when the objectives of stakeholders overlap in meaningful ways, despite the diverse meanings that different groups of actors attribute to territory and the myriad values they derive from it. In Randilen, institutions for securing pasture and protecting wildlife seem to coexist, given shared interests in preventing land fragmentation that would reduce wildlife habitat and diminish rangeland productivity. Increased tenure security and capacity for defending farms from dispersing elephants have ensured buy-in from cultivators as well. It is my emergent view that in the case of Randilen, the middle ground between stakeholders is greater than the sum of differences, and this has led to a significant rise in local support for the WMA.

Land, Institutions, and Wildlife

There are three central themes that run through this book. The first is *land*. Land underlies this study, literally and figuratively. By *land*, I am to referring to material distributions of natural resources, wildlife habitat, and pasture, but also to rights claims, tenure systems, and use patterns. The second is *institutions*. Here, I am referring to organizations, stakeholder dynamics, formal and informal governance mechanisms, management practices, norms, laws, and customs. Institutions form a layered set of multi-scalar influences that bear on community-based conservation, including the activities of international conservation foundations, local NGOs, villages, and the everyday practices of community members on the ground (cf. Lesorogol 2022). My thinking follows loosely from the work of North (1991), who conceptualizes institutions in cultural terms as "rules of the game," and of Ostrom (1990), who focuses on the various forms of social organization that facilitate management of common pool resources. The third pillar is *wildlife*. I conceptualize wildlife as a key source of tourism revenue in Tanzania, but also in terms of its livelihood costs for rural communities. My training as an anthropologist caters more toward understanding the human dimensions of wildlife conservation but less to the direct study of wildlife biology. I have endeavored in this book to weave a narrative together based on my readings, interviews, and engagements with people who generally have far more knowledge of wildlife dynamics than I do. In keeping with the interests of the pastoralists with whom I work, I also think of wildlife and livestock in a relational sense, given their competitive yet complementary roles in the ecology of rangelands (Reid 2012). I approach these three pillars—land, institutions, and wildlife—through the prisms of pastoralism and conservation, two ways of thinking that have key ideological differences but which share much common ground. My interest in this book is in highlighting some of these commonalities. While the divergences between the two epistemologies are well established (Goldman 2020, 2003), some scholars have also written about their potential marriage (Godfrey 2018). Few studies in East Africa, however, have empirically shown how the interests of conservationists, investors, government officials, and pastoralists can be brought together in practice.

My positionality as a researcher includes personal feelings and inclinations that are worth prefacing here for transparency. It is my view that wildlife conservation in Tanzania is complex and multidimensional. Critical social science can offer important insights into the perspectives and lived experiences of people who would otherwise be marginalized in conservation discourse. Fortress models of conservation that displace and dispossess local communities in favor of "pristine" wilderness spaces are an ever-present concern for people living in wildlife-rich areas of Tanzania (Brockington 2002; Neumann 1998). In my view, fortress conservation is both ethically wrong given its human cost and unproductive in the context of semi-arid rangeland ecosystems where wildlife and pastoralists must move across large areas to access resources that vary across space and time. I consider my approach to be applied in its attempt to foster long-term solutions to environmental issues in this area that are mutually beneficial for people and wildlife, and theoretical in its contribution to critical scholarship on conservation. My heart is with the plight of wildlife, and my intellectual bias is toward safeguarding community livelihoods. My intention with this book is to offer an ethnographic approach to understanding why community members feel the way they do about Randilen WMA, and perhaps in so doing, contributing to the design and implementation of sustainable and equitable conservation initiatives in the future. I do not consider myself an activist, but what is intriguing about this case is the extent to which the fate of wildlife is bound together with the state of pastoral land tenure and institutions for managing rangelands. As I will attempt to show, securing rangelands and protecting wildlife habitat are not antithetical propositions. As evidenced by a growing body of literature on the beneficial role of domestic herds in managing semi-arid rangelands, people, livestock, and wildlife can and do coexist (Homewood and Rodgers 1991; Reid 2012). I am convinced that thinking about them together represents the future of conservation in this region.

In keeping with a political ecology approach, I theorize the environment as a political arena, within which different social actors with varying degrees of power are engaged in struggles over access to, and control of, resources (Bryant and Bailey 1997; Jones 2006; Vaccaro et al. 2013). I also draw from literature in critical development studies, a field that could perhaps be distilled down to the study of social change. More precisely, development theorists consider the ways in which diverse actors mobilize their own logics, interests, values, and strategies to rationalize the trade-offs of change (de Sardan 2005). Different ways of approaching natural resources can create tensions as distinct types of knowledge come into conflict. Top-down political interventions, whether originating from colonial administrators, NGOs, or the state, often fail to accomplish their objectives as a consequence of these differences (Ferguson 1994; Havnevik 1993; Hodgson 2001; Li 2007; Scott 1998).

While the shortcomings of development are well documented, I am more intrigued by literature that highlights the potential for conservation to unify the interests of groups with independent stakes in the governance and management of natural resources. Extrapolating from Foucault's (1976, 1978) notion of governmentality, Agrawal (2005) developed the concept of environmentality in the context of community-based conservation. Governmentality refers to the tools and techniques that governments use to make people and resources governable without needing to resort to direct force (Lemke 2001, 2002). Foucault (1978) described these practices as the "conduct of conduct," or the general management of people's behaviors. Agrawal (2005) posited that through decentralization of governance institutions, individuals come to participate more "intimately" in government and realize the benefits of conserving resources, leading them to become "environmental subjects" whose interests align with the state (Agrawal 2005, 1). Community members thus participate in the production of their own governability in a Foucauldian sense by self-regulating their environmental practices, eliminating the need for the state to directly exercise power over their behavior. Other scholars, like Fletcher (2017), have since expanded upon Agrawal's theorizations of environmentality to include a diverse set of related processes and pathways through which power is exercised and subjects are made (i.e., multiple environmentalities). Though a compelling theoretical concept, environmentality runs the risk of reducing the interests and practices of rural communities to products of political processes. Political context surely plays a significant role in shaping people's attitudes toward conservation, but preexisting cultural values of stewardship often run deeper than political subjectivities (Cepek 2011). Furthermore, while people may share common ground on a particular subject, it does not necessarily mean that they think the same way and care about the same things. In short, political environments do not function independently of sociocultural context (Singh 2013).

I interpret community attitudes toward Randilen WMA through a cultural politics lens following Gardner's (2016) discussion of conservation dynamics in the Serengeti ecosystem and Hodgson's (2001) work on Maasai notions of ethnicity and gender in the context of colonial and socialist development policies. Cultural politics refer in a sense to how people conceive of themselves in relation to others (identity) and how groups of people derive material value from, and attribute symbolic significance to, space (place) (Moore 1998). Such processes of meaning-making can be myriad and contested. Here, I am particularly interested in the question of how people think and feel about a WMA, and how these appraisals reflect expressions of their own identities and life aspirations. I have found particularly fruitful de Sardan's (2005, 144–145) concepts of "sidetracking" and "selective adoption," applied when communities choose to support some elements of conservation that align with their own interests while excluding or appropriating others to fit their personal goals. Cleaver (2012) refers to this process of utilizing available institutions and refashioning them to fit a particular objective as "institutional bricolage." These concepts fit quite nicely with what I observed during my fieldwork, as community members seemed to interpret conservation in terms of their own livelihood concerns rather than subscribe to the same set of priorities as conservationists, ecologists, tour operators, or the state.

To understand local sociopolitical dynamics, I draw from literature on the human dimensions of conservation that operationalizes the concepts of environmental governance and management (Bennett et al. 2017; Bennett and Dearden 2014a). I take governance to refer to the mechanisms and pathways through which conservation-related decisions are made (Bennett 2015). Governance analyses require one to consider who the relevant stakeholders are across scales, what their roles are in decision-making processes, and how power is distributed across these actors (Bennett and Satterfield 2018). Governance processes determine what the specific management priorities of a given conservation area will entail in practice, but they are not synonymous with management. An example of a governance process is holding a steering meeting to discuss future strategic planning. I understand environmental management as the institutional arrangements that are put into practice on the ground to regulate and enforce the outcomes of governance decisions (Bennett 2015; Bennett and Dearden 2014a). Examples of management could include carrying out ranger patrols, issuing fines, confiscating livestock, or educating people through rural outreach programs. Governance and management come hand in hand, but for the sake of analytical clarity, it is productive to parse them. Participatory governance processes can build trust and respect and ensure that local livelihood concerns are prioritized in the conservation model. Similarly, management institutions that integrate the traditional values of community members are more likely to cultivate support for conservation. By contrast, exclusionary processes of decision-making taken in a top-down fashion by authorities can make people feel marginalized and resentful (Bennett and Dearden 2014b). These dynamics become particularly impactful when coupled with stronghanded management practices, which can engender the perception that conservation is a tool for policing the community. Thus, governance and management are useful concepts for understanding why community members support or contest a particular conservation area.

Maasailand and the Tarangire Ecosystem

The regional context of this book is Maasailand, a semi-arid landscape spanning roughly 150,000 km² across southern Kenya and northern Tanzania. Maasailand is revered globally for its vast savannas, charismatic wildlife, and iconic pastoralists. The majority of people living in Maasailand are Maa-speaking and self-identify as Maasai (Homewood and Thompson 2010), though ethnicity is politicized and sometimes contested (Gardner 2016). Most Maasai are herders and cultivators whose livelihoods depend on livestock (Homewood et al. 2009). The Tanzanian portion of this landscape covers about 70,000 km² and was part of a single administrative district until 1974. It has since been divided into five smaller ones: Monduli, Ngorongoro, Kiteto, Simanjiro, and Longido. Kiteto and Simanjiro are located within Tanzania's Manyara Region, and the rest are part of the Arusha Region.

Numerous works describe Maasai culture in depth already (cf. Spear and Waller 1993), so I provide some succinct ethnographic context on their way of life here. The Maasai traditionally organize themselves into territorial sections called *iloshon* (singular *olosho*), each functioning as its own self-contained unit and as a part of a larger whole (Ndagala 1997). The member villages of Randilen WMA are inhabited mainly by Kisongo Maasai pastoralists (*IlKisongo*) and Arusha Maasai agropastoralists (*IlArus* or *WaArusha*). According to oral histories from Maasai elders, the Kisongo have been present in the area for about two hundred years following their southern expansion from Kenya into Tanzania sometime between the sixteenth and eighteenth centuries (Galaty 1993; Hodgson 2001). The Kisongo view themselves as "people of cattle," symbolizing the significance of livestock to their social and economic life (Galaty 1982). They maintain a pastoral mode of production supplemented by subsistence cultivation of maize and beans, though they have diversified their economy over the past forty years to include wage labor and business activities (McCabe et al. 2010; Trench et al. 2009). Like other Maasai sections, the Kisongo value collective management of shared pastures through common property mechanisms and territoriality at the nested scales of section, subsection, clan, and family. They herd cattle, sheep, and goats through an assemblage of customary institutions, including a Nilo-Hamitic

A Kisongo Maasai woman smiles while sitting in her home in Lemooti village. Photo by author in 2019.

An Arusha man in Lolkisale village poses with his herd of goats and sheep. Photo by author in 2019.

age-set system, a gender-based division of labor, and a decentralized political system (Fosbrooke 1956b).

The Arusha speak a close dialect of Maa and share numerous social institutions with the Kisongo, including the same age-set system (Spear and Nurse 1992). Like the Kisongo, the Arusha greatly value livestock for the social capital they generate in Maasai society and as economic stores of wealth across a semi-arid landscape. The core of the Arusha economy, however, is cultivation of beans, maize, peas, and other crops like sunflowers grown mainly for commercial distribution. The Arusha generally prefer individual private property and sedentarized houses constructed from modern building materials, as compared to the semi-nomadic mud huts inhabited by the Kisongo. In areas where Arusha and Kisongo live in close proximity to each other, visible differences between the groups are less distinguishable than in other contexts, though social distinctions endure.

Competing theories exist as to the ancestral lineage of the Arusha in relation to the wider genealogy of Maa-speakers in eastern Africa. The Arusha point to the confluence of Pangani River just south of Moshi town as their ancestral homeland, but their exact origins have been the subject of scholarly debate (Fosbrooke 1948, 1956a; Spear and Nurse 1992). The Arusha likely originated from historical interactions between Pare, *IlKikoin*, *Iloogalala*, and

Maasai people that included warfare and economic exchange with the Kisongo in the precolonial era (Fosbrooke 1948; Kuney 1994; Spear 1997). What is known for sure is that the Arusha first took up residence on the fertile southwestern slopes of Mount Meru around 1830, where they have practiced intensified cultivation since. Oral life histories from Arusha elders living in Randilen's member villages reveal that the Arusha began settling the area in the 1950s during a period of accelerating land scarcity on the slopes of Mount Meru (Bluwstein 2017; Igoe 2010). The Arusha leveraged cultural, economic, and social connections with the Kisongo to expand into pastoral areas during the late colonial and socialist periods (Hodgson 2001; Kuney 1994). They are now the dominant ethnic group across the WMA's member villages (Raycraft 2022a).

The Arusha and Kisongo are closely tied in this ethnographic context, with families drawn together through shared ceremonies, rituals, and patterns of intermarriage and exchange. It would thus be reasonable to classify both groups as Maasai sections, a stance consistently taken up by Spear (1997) and echoed by Randilen's community members themselves. At the same time, ethnic territoriality is notably pronounced across Randilen WMA member villages and sub-villages, and individuals are quick to point out distinctions between the two groups. It would thus also be fair to consider them different ethnic groups with distinct yet overlapping interests. Underlying the notable commonalities between the Kisongo and Arusha lie differences in the ways in which each group thinks about land and resources. When asked how they conceive of their identities and livelihoods, the Arusha make clear that they consider themselves farmers who keep livestock rather than herders who farm, a key enduring distinction between the Arusha and the Kisongo residing in Randilen's member villages. Interested readers could refer here to the work of Thomas Spear (1997) for further context on the ethnographic textures of Arusha and Maasai ethnicity.

Randilen WMA is located in Monduli District in the Tarangire ecosystem, a landscape dominated by *Commiphora* bushland, *Acacia* and *Vachellia* woodlands, and grassland plains that supports a diverse assemblage of wild ungulates and large carnivores (Kiffner et al. 2022; Kissui et al. 2022). The central part of the ecosystem, where Randilen WMA is located, is home to more than five thousand elephants, the largest subpopulation in northern Tanzania and one of the fastest growing continent-wide (Foley and Foley 2022). In total, the ecosystem spans approximately 25,000 km^2 and encompasses both Tarangire NP and Lake Manyara NP as well as several other community-based conservation areas (Kent and Litchenfeld 2024). The area is inhabited by Kisongo Maasai pastoralists, Arusha Maasai agropastoralists, and several other ethnic groups and is accordingly referred to as the Maasai Steppe in some discourses.

Rainfall varies significantly in the Tarangire ecosystem and generally follows a bimodal pattern, with a mean annual amount of about 656 mm (Foley and Faust 2010). The lowland plains receive as little as 250 mm of annual rain, while some mountains get upwards of 1,500 mm (Ndagala 1994). Much of the landscape gets less than 500 mm on average, with vegetation growth limited to the rainy periods (MDC 1994, 5). The ecosystem undergoes considerable transformation across seasons. In the long dry season, from approximately June to October, wildlife congregates around the Tarangire River inside Tarangire NP. The months of November to January usually coincide with a short wet season, followed by a short dry season from January to February. In the long wet season, from February to May, grasslands become verdant, and wildlife disperses outside the park (Sachedina and Nelson 2010). Tarangire NP thus functions primarily as a dry season park (Borner 1985). The dynamic nature of this ecosystem makes it particularly challenging to protect through conservation initiatives (Bond et al. 2022). Seasonal dispersals outside Tarangire NP leave wildlife vulnerable to competing land uses and often lead to negative interactions with local people who depend on their farms and livestock for livelihood (Hariohay and Røskaft 2015; Raycraft 2024a, 2024b; Raycraft et al. 2024). This issue has generated considerable interest from conservationists hoping to protect wildlife habitat areas adjacent to Tarangire NP that overlap with village land (Kiffner et al. 2024). Spearheaded by the AWF and other NGOs, various community-based conservation initiatives have been implemented over the past thirty years in an effort to prevent these areas from becoming fragmented. Randilen WMA is one such initiative and is the primary focus of this book.

Methodology

The empirical foundation of this book is eighteen months of ethnographic field research (July 2019–July 2020; June–July 2022; April–May 2023; April–May 2024) contextualized through historical analysis. The book draws from extensive participant observation of everyday life across Randilen's member villages and three hundred interviews with key stakeholders with pertinent knowledge about Tanzania's wildlife conservation sector. I spent a significant portion of my time in the field attending conservation governance meetings at different scales, tagging along on ranger patrols, herding livestock, telling stories by the

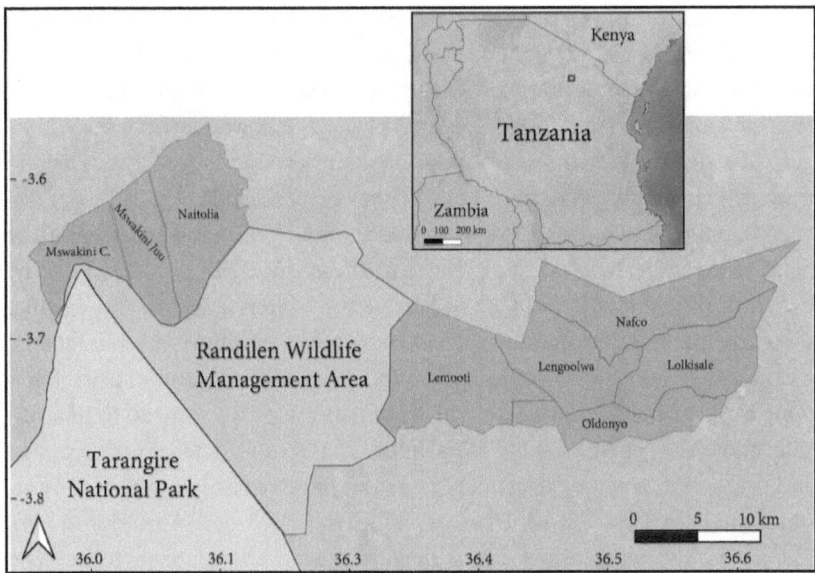

Location of Randilen Wildlife Management Area in the Tarangire ecosystem including the approximate boundaries of its eight member villages. The smaller square map at the top shows Randilen's regional location in northern Tanzania.

fire, hiking through village land with local guides ("walkabouts"), collecting and photocopying documents from village offices, and observing cattle-exchange ceremonies and rituals. The narrative I present about Randilen WMA is grounded in the lived experiences of local community members elicited through 160 in-depth qualitative interviews with residents of Randilen's eight member villages (Naitolia, Mswakini Juu, Mswakini Chini, Lemooti, Lengoolwa, Nafco, Lolkisale, and Oldonyo), and 40 interviews carried out across two other nearby villages in the Tarangire ecosystem—Olasiti, which neighbors Mswakini Chini to the southwest and Makuyuni, which borders Naitolia to the northeast. The story of Randilen WMA begins in Lolkisale village (chapter 2).

In 2019–2020, I conducted interviews with ten men and ten women in each study village at people's homes. Bomas (pastoral homesteads) were selected with a goal of producing a sample that was well distributed geographically. In Tanzania, villages are divided into sub-village hamlets called *vitongoji*. Interviewees were recruited from across *vitongoji*, with recruitment numbers depending on sub-village population size. I selected homesteads myself in a semi-random fashion until participant quotas (by gender and village) were reached. The interviewees who chose to participate were well represented in terms of age, gender, and socioeconomic background. That said, the intention was not for these interviews to serve as a representative sample on their own, but to generate qualitative themes and help conjure a picture of how people felt about the WMA. Interviews spanned about an hour and were audio-recorded and later transcribed. Edwin Maingo Ole, a wonderfully skilled linguist, assisted me with translations during interviews, and Elsa DeLuca helped transcribe the audio recordings upon my return to Canada.

While the qualitative interviews across the study villages were essential for helping me understand local perspectives about the WMA, I was puzzled by the stark differences I was finding between my interview data and existing scholarly literature on Randilen WMA, which framed the WMA as a fortress model of conservation that excluded and dispossessed communities (Brehony et al. 2018; Loveless 2014). This dissonance motivated me to think critically about the generalizability of my findings. I became determined to design and implement a survey to assess, in a representative fashion, community attitudes toward Randilen WMA across all member villages (April–July 2020). The instrument was close-ended, with coded numerical responses to facilitate data entry and analysis, and administration took about half an hour. Using yes-no responses, three-point scales, and five-point Likert ordinal scales, the questionnaire systematically examined key metrics like attitudes toward the WMA, trust in conservation authorities, perceptions of costs and benefits, lived experiences of governance and management, and evaluations of the WMA as a success or a failure. Rather than something that stands alone, the survey serves to triangulate my ethnographic findings and qualitative data.

To determine the study population, I worked with sub-village chairs who traveled to each boma in their jurisdictions on motorbikes to compile complete lists of household heads in each of Randilen's sub-villages. I then established two sampling frames—male household heads and female household heads—and used Cochran's (1963) formula for calculating sample sizes for finite populations, using a proportional value of 0.5, a 95 percent confidence interval, and a 5 percent margin of error to determine representative samples. To select household heads, I employed stratified random sampling using sub-villages as strata, and I selected samples from each that were proportionately weighted based on sub-village population sizes relative to the total frames. Once I had determined sample sizes for each sub-village, I numbered the household heads from each frame and used a random number generator to select the numbered household heads until the quotas for each stratum and frame were reached. With the assistance of local field assistants, I administered the survey to 326 male heads and 191 female heads. To address the significantly higher number of male heads, a trend in line with cultural norms in

Maasai society, I decided to establish a third sampling frame comprising females in male-headed households. This meant administering the survey to the senior wife of every second surveyed man, which amounted to an additional 161 women. Thus, the total number of people surveyed was 678 individuals.

The combination of participant observation, surveys, and interviews helped produce a triangulated ethnographic account of Randilen informed by the perspectives of community members and other conservation stakeholders. Based on this research, I present here an anthropological narrative of Randilen's rise to acclaim, beginning in the next chapter with an overview of the historical context that frames my analysis.

CHAPTER ONE

Maasai Society and the State

A history of layered land policies shapes the political landscape of Tanzania's Maasailand and underlies current tensions between state authority and the self-determination of pastoral communities. German colonial rule (1890–1918) marked the beginning of centralized land and wildlife conservation policies in Tanzania (formerly Tanganyika). Under the German Imperial Decree of November 26, 1895, all land within the jurisdiction of German East Africa was classified as "unowned" and placed under the control of the Reich (Fimbo 1992, 156). German administrators subsequently established a series of game reserves for regulating hunting and reallocated millions of acres for commercial cash crop plantations and settler farms, undercutting existing African customary land claims (Fimbo 1992; Shivji 1998).

Following the First World War, Britain took over governance of Tanganyika in 1919. The British administration built on the German groundwork of centralized resource control through a series of laws, including the Land Ordinance of 1923, which declared all land in Tanganyika as property of the crown (Shivji 1998). Game Ordinances in 1921, 1940, and 1951 and National Park Ordinances in 1948 and 1959 paved the way for the formalization of an expanding network of game reserves and national parks, establishing in the process the paradigm of fortress conservation that is still visible in contemporary Tanzania (Brockington 2008). This collection of protected areas included the Ngorongoro Crater, initially grouped together with the Serengeti plains following the Game Ordinance of 1940, before being parsed into the Ngorongoro Conservation Area (NCA) and Serengeti National Park in 1959. Large-scale allocation of productive lands to European farmers deepened the effects of the enclosures, as did the creation of the Masai Reserve in 1922 (renamed Masai District in 1926), which restricted the reciprocal networks and herding mobility of the Maasai and required them to reside in a fixed territory (Hodgson 2001).

After independence, President Nyerere was elected in 1962 and forwarded his vision for African socialism. He mobilized the rhetoric of *ujamaa* to create a sense of "extended family" shared by all Tanzanians regardless of their ethnicity, cultural traditions, economic practices, or geographical areas of residence. Nyerere outlined his model of national development through the Arusha Declaration of February 5, 1967, which proposed to draw the nation together through a one-party political system, compulsory formal education, nationalization of key sectors, collectivization of agricultural production, and an ethos of self-reliance from the influences of imperial powers (Huizer 1973; Nyerere 1968). Based on suggestions from the International Bank for Reconstruction and Development, Tanzania pursued rural agricultural development policies rather than urban industrialization, leading to the formation of farming cooperatives and partnership ranching associations (Hodgson 2001). Maasailand played host to several major initiatives during this period, including water development projects and a major USAID-funded ranching program aimed at "modernizing" pastoralists by transforming them into commercial beef producers. Ultimately, most of these initiatives failed to accomplish their goals due largely to a lack of appreciation for the enduring effectiveness of customary pastoral production systems (Gardner 2007; Hodgson 2001; Parkipuny 1979).

Following the Arusha Declaration, President Nyerere rolled out the *ujamaa vijijini* ("neighborhood villagization") initiative through which the government encouraged rural communities to reorganize their labor arrangements into collectives. *Ujamaa vijijini* was meant to be a democratic process built on trust, whereby rural communities chose to participate in the program of their own accord (Shivji 1986). However, what began with socialist ideals for cooperative economics ultimately gave way to a coercive model of resettlement (Shivji 1986), referred to in Maasailand as "Operation *imparnati*" ("permanent habitation" in Maa) or villagization more generally (Ndagala 1982, 29). Contrary to the prior approach, where people were encouraged to voluntarily move to cooperative *ujamaa* villages, Operation *imparnati* was carried out with the use of force. In 1971, initial resettlements were implemented in the Dodoma region, and following their swift execution in the face of general resistance, the state determined in 1974 that all rural Tanzanians had to live in villages (Kjekshus 1977; Scott 1998). This set the stage for the National Villages and Ujamaa Villages Act of 1975, which established villages as the primary unit of settlement in rural Tanzania and created a framework for them to be registered and titled by the state. The Tanzanian government subsequently carried out one of the largest resettlements of people in history, with approximately five million people forcibly relocated (Scott 1998).

Villagization disregarded seasonality and cultural adaptations to challenging environments, including mobility and improvisation (Scott 1998; Shao 1986). While the resettlements were associated with development promises, most of these services failed to materialize. Resistance was met with violence, and in some cases, people's homes were burned. While one of the objectives of villagization was to streamline production, Scott (1998, 224) astutely points out that "the thinly veiled subtext of villagization was also to reorganize human communities in order to make them better objects of political control." Put differently, the process was meant to eliminate ethnic, social, and class-based differences and enforce a model of egalitarian citizenship.

In Maasailand, these changes disrupted the stability of the pastoral system by further reducing the total amount of productive land available to herders and creating new boundaries limiting access (Kuney, personal communication, 2020). Villagization created autonomous localities, each with separate jurisdictions of authority and land use planning systems. Within these "governable spaces," land could be allocated to individuals and families through the formal apparatus of state legislature, irrespective of preexisting pastoral institutions for managing rangelands (Watts 2004, 50). The lasting effects of villagization in Maasailand were not necessarily the physical reorganization of villages, which did not significantly alter pastoral settlement patterns (Homewood and Rodgers 1991), but rather the institutional shifts in governing and managing rangelands (Gardner 2012, 2016; McCabe et al. 2020). Legally, villages supersede the traditional territorial arrangements (sections, subsections, clans, and families) of the Maasai for determining acceptable land uses based on customary systems of governance. Villages still exist in Tanzania as administrative units that enable local government councils to allocate land within their jurisdictions (LaRocque 2006). Thus, the enduring outcome of villagization in Maasailand has been the fragmentation of the region's political landscape into localities.

The social implications of villagization in Maasailand were further complicated by the continued development of the wildlife sector. The passing of the Wildlife Conservation Act of 1974 mirrored colonial game laws from decades prior. Rather than revitalize customary land rights and Indigenous stewardship of ecosystems, the policy once again recentralized resource control in the post-independence era by formalizing all protected areas as state property. National parks and the NCA were designated exclusively for wildlife preservation, and game reserves, Game Controlled Areas (GCAs), and open areas were allocated as state-managed trophy hunting blocks. While game reserves prohibited local settlement, GCAs and open areas often overlapped village land, creating a pluralistic legal framework for land governance characterized by institutional layering.

While President Nyerere had been steadfast in his promotion of self-reliance in the face of international influence, his successor, President Mwinyi, adopted a different tone. Following his election in 1985, President Mwinyi signed on to a structural adjustment program with the World Bank and International Monetary Fund that overhauled the national economy through neoliberal reforms aimed at stimulating a new trajectory of capitalist development. In general, the International Monetary Fund and World Bank used a series of strategies to boost economic growth in developing countries. Some of the key adjustments included devaluing currencies relative to the U.S. dollar, trimming subsidies, eliminating price controls, privatizing sectors that were previously public, liberalizing international trade, encouraging foreign investment, and introducing multiparty political systems (Agrawal et al. 1993). In Tanzania, a number of policies were put in place: controls on crop prices were lifted, the shilling was devalued, food trade was liberalized, and public expenditure was limited as part of a tighter monetary policy (Agrawal et al. 1993; Gibbon and Raikes 1995). In a nutshell, the adjustments were meant to encourage the adoption of free-market economic principles that promoted competition and privatization, as well as a "rolled back" state that did not protect against foreign investment.

The consequences of Tanzania's structural adjustment program have been actively debated by scholars and practitioners. Some in the conservation field point out that state-private partnerships often neglect the rights of resource-dependent communities and favor actors with greater access to capital (Igoe and Brockington 2007; Igoe and Croucher 2007). By contrast, Gardner (2016) demonstrates that the reforms also generated meaningful opportunities for pastoral communities to partner directly with tourism investors in the early 1990s. Certainly, the question of whether "neoliberal conservation" does more harm than good is a contentious one in both the global and regional literature (Büscher et al. 2012; Fletcher et al. 2014). Complex discussion aside for now, it is difficult to decipher whether neoliberalism has endured in Tanzania, as the state still plays a dominant role in regulating private operations that affect its ability to extract revenue from national resources (Gardner 2016). Lasting tensions between centralization and neoliberalization in contemporary Tanzania seem to suggest a blend of socialism and capitalism rooted in a long history of institutional layering (cf. Barkan 1994).

A Struggle for Sovereignty

By the 1980s, the cumulative effects of colonialism and socialism on Tanzania's Maasailand had resulted in the alienation of pastoral lands and significant constraints on customary patterns of rangeland management. In the face of ongoing threats to their way of life, Maasai pastoralists have since adapted to the changing political environments of the modern nation state by learning to wield state institutions in ways that forward their own interest in protecting the livestock economy. When the state pressures pastoralists from above, herders push back from below. State power and bottom-up resistance are thus two sides of the same coin, and it is shortsighted to focus on one without the other. Rather, a political ecology approach contextualizes the dynamics between Maasai society and the Tanzanian state as an ongoing *struggle* over territory and resources. In instances where the state has implemented top-down policies, the Maasai have exercised their agency by refashioning them in support of pastoralism. When local mobilization threatens state authority, the central government amends its laws to reconsolidate power. Legislative reforms at once create new threats and also new tools for the Maasai to use in ways that challenge state oppression and reclaim autonomy over pastoral territories. Layered policies for governing land and wildlife create a formal arena for mobilization of state and community interests while also creating an ambiguous playing field demanding legal dexterity to maneuver. Ever at stake for Maasai communities and the state is *sovereignty*, defined here as the power to take and enforce governance decisions and exercise authority over territory.

Villages offer one example among many of herders working within the constraints of formal policy to exercise their agency in relation to the state. Though villagization was implemented in a top-down fashion during the socialist period, Maasai leaders realized following structural adjustments that villages also represented institutions that could be used to their advantage. Lazaro Parkipuny, the first Maasai person to receive an advanced degree and the founder of the first pastoral land rights nongovernmental organization (NGO) in Tanzania, led the way. Parkipuny's organization, called the Korongoro Integrated Peoples Oriented to Conservation (KIPOC—meaning "we will recover" in Maa), recognized that the pastoral economy was becoming constrained by land alienation and sought to use the village as a means of securing common pastures. Villages—once a state strategy for reordering landscapes to make rural communities more "legible" and governable—became a way for pastoralists to undermine the ability of the central government to reallocate pastoral land to parastatals or lease it out to private companies (Scott 1998).

Parkipuny's struggle against Tanzania Breweries Limited in Loliondo exemplified these tensions. In the late 1980s, Tanzania Breweries proposed to establish a commercial barley and wheat farm spanning 100,000 acres of crucial grazing area (see Gardner 2007, 114). Realizing that vital rangelands were at risk of being enclosed, Parkipuny and the local Maasai Member of Parliament swiftly rallied villages in opposition against the grab (Gardner 2007). Parkipuny held that villages should be able to exercise customary rights to land, owing to the history of villagization. Titling villages thus became of crucial importance to pastoralists because it provided a means of formalizing customary claims. In defense of pastoral land, Parkipuny pushed villages to register and title their boundaries pursuant to the National Villages Land Act. The main argument put forth by Parkipuny's followers was that village councils should have sovereignty to grant concessions on their land, provided that villages had been formally registered (Gardner 2007, 114). Villages thus created legal grounds for pastoralists to push back against dispossession in the neoliberal era and protect the pastoral way of life.

The conflict in Loliondo between villages and state was symbolic of a larger tension in Maasailand over the rights to govern rangelands. Despite Parkipuny's headway, the state remained intent on maintaining authority over national resources and viewed customary rights as unnecessary competition (Homewood and Thompson 2010). In response to the rising power of villages over valuable lands, the state reformed its land laws in the late 1990s to ensure that the category of village land only referred to areas where communities could demonstrate that they were using land productively. In the eyes of the state, "productive use" was defined in terms of sedentarized settlements and crop cultivation. Though integral to the pastoral economy, and beneficial for preventing fragmentation of wildlife habitat adjacent to national parks, common pastures used by herders on a seasonal basis were vulnerable to being labeled by the state as unproductive and outside the auspices of villages. Two acts passed in the same year—the Land Act No. 4 of 1999 and the Village Land Act—codified the potential for state reappropriation of village land. The new acts divided all land in Tanzania into three categories: village land, reserve land, and general land. Accordingly, public lands were defined as general land unless they existed inside a protected area or a village. Land that was directly used by villagers was considered village land, but parts of villages that the state viewed as either underutilized or "unoccupied" were at risk of being reclassified as general land (Gardner 2016, 55; Shivji 1998). In the early 2000s, Maasai leaders in Loliondo were particularly concerned that the new stipulations would allow the state to grab back lands that pastoralists had established control over using the village institution. Importantly, village titles secured prior to the new acts were made obsolete by the reforms, forcing villages to go through new legal processes of demarcating village boundaries (see Gardner 2016, 175).

Though the new acts threatened to undermine pastoral tenure, they simultaneously created a new formal framework for articulating rights claims. Previously, the National Villages Land Act had provided a means for pastoralists to demonstrate customary claims to land by titling villages; however, it did not specify *how* customary claims within villages translated into formal tenure. The revised acts of 1999 stipulated that villagers had customary occupancy rights on village land that could be represented formally through Certificates of Customary Rights of Occupancy (CCROs). CCROs essentially codified customary tenure within villages, though their designation as occupancy rights as opposed to ownership highlighted the fact that all land in Tanzania technically remained state property after the land reforms. Though less secure than individual title deeds, CCROs were backed by state law, making them appealing options for pastoralists struggling to articulate their land rights. In quick order, Maasai communities adapted to the new political landscape by re-titling villages and obtaining CCROs, which they reasoned would offer greater tenure security than before and strengthen the autonomy of villages to govern territory. Pastoral land rights NGOs, like the Ujamaa Community Resource Team (UCRT), assisted with these processes, which required intimate knowledge of the legal system in Tanzania. To obtain CCROs, villages first needed to establish clear boundaries before applying for Certificates of Village Land to be issued by the district government along with vector maps. Villages then needed to be officially registered as village land on the national registrar of villages—a fundamentally political undertaking in Tanzania given the high stakes involved. Once villages were registered, they were required to carry out land use planning and establish bylaws that were agreed upon by both village and district government. At that point, village councils gained the formal power to allocate land within their jurisdictions and provide individual villagers with CCROs.

Maasai leaders were hopeful that formalization of villages under the new acts would give them the tools needed to prevent pastoral land from being grabbed by the state or private companies. At the same time, however, they were wary of the potential of individualized tenure claims to undermine customary forms of communal rangeland management. Rapid subdivision of Maasai group ranches in southern Kenya in the late 1980s provided striking examples of the potential for individualization to fragment the pastoral commons (Galaty 1994, 118). To function effectively, the Maasai herding system requires collective institutions to regulate shared access to and use of rangelands. Individual tenure claims could thus undermine the pastoral mode of production by incentivizing people to establish exclusion criteria for their own personal plots of land. Despite the benefits that village titling afforded Maasai communities, disaggregation of village land into individual CCROs led in some cases to rangeland fragmentation in the form of expanding smallholder farming. This trend became particularly pronounced across the Arusha-inhabited villages in rural Monduli district, where Randilen WMA is located.

Historical patterns of Arusha in-migration to Kisongo territories in Monduli district since villagization epitomized the challenge of using CCROs to defend pastoral land in multi-ethnic villages. Linguistic, cultural, and economic proximity to the Kisongo allowed the Arusha to move into pastoral areas informally by presenting themselves flexibly as "Maasai" (Kuney 1994). Once established in rural Monduli, the Arusha won positions on village councils and used local government institutions to allocate land to other Arusha along kinship lines, accelerating patterns of in-migration and increasing their political representation and territorial control. Fundamental differences in mode of production—with Arusha preferring crop cultivation and individual private property, and Kisongo prioritizing herding on shared pastures—meant that villages quickly became susceptible to fragmentation and conflict, depending on ethnic dimensions of local governance. These dynamics are particularly entrenched in the Randilen WMA member villages (chapters 3 and 4).

The pitfalls of individualized land tenure in pastoral areas gave rise to the concept of communal CCROs, conceptualized by UCRT to ensure that pastoralists could secure land collectively. Communal CCROs are obtained through the same legal process, but they are allocated to groups rather than individuals. Maasai leaders hoped that communal CCROs would provide secure tenure for communities that relied on shared institutions for accessing and using resources across seasons. Through dedicated efforts, UCRT managed to secure more than 100,000 acres of communal CCROs across the Simanjiro plains east of Tarangire National Park for the Maasai, as well as 50,000 acres for the hunter-gatherer Hadzabe community living near Lake Eyasi.

Communal CCROs have been vital for securing pastoral tenure across northern Tanzania, but they also face some real limitations. For example, CCROs are restricted to a maximum of 250 hectares, as this is the maximum that villages are allowed to allocate. Registering larger areas requires explicit consent from the Commissioner of Land and must be mapped and demarcated through a District Authority Review. Furthermore, although CCROs signify formal tenure status on paper, they do not offer an institutional framework for management in practice. Dynamics on the ground are shaped by the rights holders based on existing property relations and agreements around the rights and responsibilities of resource users. In instances where the community collectively implements a land use plan that stipulates local grazing bylaws, areas under the jurisdiction of communal CCROs are generally managed effectively. However, CCROs refer only to tenure status and do not, in and of themselves, translate into rangeland management practices. Of particular consequence, CCROs do not directly increase the capacity of rights holders to enforce bylaws within their territories. This makes it challenging for pastoral communities to physically enforce boundaries on communal grazing areas when facing encroachments from outsiders. Enforcing CCROs is especially challenging when domestic elites or politicians are involved and are in search of pastures to graze their personal herds. CCROs make clear that such transgressions are a violation of the formal classification of the land. However, material defense of territory is contingent upon the presence of other institutions on the ground.

Perhaps most significantly, the very thing that CCROs aim to secure—pastoral tenure—can be made insecure by the state in areas coveted by the central government for wildlife tourism. Historical discontinuities between land and wildlife policies in Tanzania mean that villages and the state can lay competing claims to territory and resources. While CCROs provide tenure security on paper, they do not apply to *wildlife*, which is owned exclusively by the state. An entrenched policy issue in Tanzania is that land is fixed, and wildlife moves across it, resulting in the institutional layering of wildlife laws on top of land acts. In practical terms, this means that in areas without abundant wildlife populations, CCROs and village titles may well hold as secure forms of tenure for pastoral communities. However, in areas near unfenced national parks with regular wildlife dispersals, CCROs come into conflict with the state's interests in wildlife.

Wildlife as Capital

The global emergence of safari tourism as a desirable recreational activity for foreigners in the late 1980s complicated the political dynamics of Maasailand. Northern Tanzania became known globally as a safari hotspot for international tourists hoping to photograph, or hunt, charismatic and endangered species of large mammals, including the well-known "big five" (elephant, rhino, lion, buffalo, and leopard). Wildlife conservation, once a colonial strategy for maintaining big game populations for settler hunters, became connected through globalization to neoliberal flows of money, people, and ideas. Wildlife was no longer just valued by the state, but also by global actors including tourists, conservation NGOs, and prospective investors looking to profit from the emerging industry. With a network of wildlife-rich protected areas already in place, Tanzania became a highly marketable destination to western tourists, and thus an attractive place to invest.

In keeping with neoliberal logics, wildlife was attributed new value as a commodity that could be "produced" through conservation initiatives and "consumed" through patterns of recreational tourism (Igoe 2017). But wild animals, in and of themselves, were not synonymous with capital. To borrow from de Soto (2000), wildlife, like all natural resources, needed to be converted into capital through political institutions for generating tourism revenue. The colonial and socialist legacy of national parks and game reserves created a viable starting point in this regard. Photographic tour operators could invest directly inside park boundaries and benefit from the unpeopled enclaves produced by the state through fortress conservation to offer safari experiences of a rugged African wilderness. For those operators looking to target international trophy hunters, they could focus their investments in game reserves, which also prohibited human settlements. Images of vast savannas and iconic wildlife could be used to market and package these experiences to potential clients from around the world (Igoe 2017). And from a government perspective, these activities were neatly manageable within a centralized financial ecosystem for collecting revenue from park fees and hunting permits.

The key governance challenge to the protected area model of conservation in Tanzania was that large mammals were not restricted to the boundaries of parks and reserves and required connected landscapes to thrive. Wildlife dispersals were particularly pronounced in the Tarangire ecosystem, where elephants and other animals regularly moved outside the park seasonally (chapter 3). Movement of wildlife across protected area boundaries meant that parks and reserves were insufficient for "unlocking" the value of wildlife in de Soto's (2000) sense.

In the late 1980s and early 1990s, tour operators began to focus their attention on the pastoral areas adjacent to national parks in Maasailand, realizing that they hosted significant wildlife populations. Villages represented appealing investment opportunities because companies were less restricted by the types of activities they could offer to tourists and less constrained by state bureaucracies regarding permits and licensing. Tourists could engage in walking safaris or enjoy "cultural" experiences by visiting curated portrayals of Maasai bomas, allowing operators to diversify their offerings. Maasai communities were often welcoming to photographic tour operators because they could negotiate the terms of investments directly through the village institution without interference from the state. Maasai leaders reasoned that if villages could be used to resist state-led land grabs, they could also be used to capitalize on dispersing wildlife in support of local development objectives. Village-based wildlife tourism, made possible by neoliberalization of the national economy, thus planted the seedlings of community-based conservation in Tanzania (Gardner 2016).

Joint ventures between tour operators and villages were structured in a variety of ways, depending on the terms of each agreement. The specific demands varied on a case-by-case basis but usually involved setting aside a concession for operators to establish lodges or camps. Inside concessions, communities were expected to refrain from clearing land for agriculture or grazing their livestock near camps, depending on seasonal availabilities of pasture. Concessions facilitated wildlife movements through village land and ensured that operators would be able to offer a safari package that mapped onto tourist expectations of "wild" Africa. In exchange, investors would agree to make infrastructure investments or cash payments to the village government to be used by the village through local governance institutions to help improve social services. Some scholars refer to this dynamic as a form of "payment for ecosystem services" (Nelson et al. 2010). Payments could entail a variable share of tourism revenue streams or fixed property rental fees, depending on the terms of the agreement. Companies often requested labor in the form of guides and security guards for camps, and negotiations were carried out as to whether the company would retain exclusive rights to tourism in the area (Wøien and Lama 1999). Once conditions were agreed upon, village councils and investors signed contracts as involved parties, which were then ratified by the central government.

Direct investments in villages in the early 1990s offered insight into the potential efficacy of neoliberal conservation in Tanzania. As discussed at length in the theoretical literature, neoliberal conservation entails a "rolled back" state and the privatization of conservation practice (Holmes and Cavanagh 2016; Igoe and Brockington 2007). In keeping with free-market capitalist principles, supply and demand dictate market conditions, and business competition is facilitated by an impartial state. Critics of neoliberal conservation argue that neoliberalization can lead to exploitation of local communities, as natural resources become commoditized and attributed new forms of economic value independent of their significance for local livelihoods (Büscher et al. 2012; Büscher and Davidov 2013; Büscher and Fletcher 2015; Fletcher et al. 2018). Indeed, the risks are significant, as in the absence of regulatory mechanisms from the state, private investors can operate in "bad faith" by thinking only about their bottom lines without catering to the concerns of local communities. At the same time, the opposite is also conceivable. A well-intentioned investor could forge a good faith relationship with a local community and support local livelihoods while generating revenue for the investor (Gardner 2016). Since both outcomes are possible, painting all private institutional arrangements as negative under the umbrella label of "neoliberal conservation" is not productive. Cases vary due to numerous contextual factors like geography, ecology, history, cultural context, and individual-level differences in investor personality.

Outcomes of the direct investment model varied across northern Tanzania, with some cases raising concerns about the potential for disrespectful or ill-intentioned investors to exploit communities (Goldman 2020; Wright 2017). Lower levels of state regulation meant that communities were potentially vulnerable to signing on to contracts that could be used to marginalize their interests over the long-term. At the same time, numerous instances of mutually enhancing examples developed in the 1990s. Particularly fruitful relationships in Loliondo, as documented by Gardner (2016) and in Lolkisale (chapter 3), made clear that the village-based model of wildlife tourism created possibilities for genuinely collaborative and community-based models of conservation (Nelson et al. 2010). Investors benefited from the arrangements, and pastoralists maintained tenure security while capitalizing on wildlife dispersals on village land. Even tourists could feel good about contributing to local development initiatives and visiting concessions that were not made at the expense of human evictions. The key stakeholder that was not benefiting from the arrangements in Loliondo and Lolkisale was the state.

What the catchphrase of neoliberal conservation struggles to encapsulate is that a neat binary between state and market almost never exists, as cases with some elements of neoliberalization are generally characterized as well by varying degrees of state protectionism, particularly when valuable resources are at stake (Holmes and Cavanagh 2016). While villages were well positioned to tap into wildlife-related capital flows in the late 1980s and early 1990s through the direct investment tourism model, the central government was not high on the prospect of competing with communities for access to the benefit streams of a resource that it viewed as state property. In fairness to the Tanzanian government, private contracts were being negotiated by investors and village councils on a case-by-case basis without a formal regulatory structure for taxation. As has become commonplace throughout Tanzania's history, the state once again reared its head by reforming national wildlife laws in the late 1990s to reconsolidate state control over the sector. To contextualize these policy changes, it is crucial to understand the macropolitical context of wildlife management in Tanzania.

Politics of Decentralization

Tanzania's wildlife sector reflects an inherent tension between neoliberalization and centralization with historically embedded roots. Decentralization of the Ministry of Natural Resources and Tourism (MNRT) into three distinct administrative branches—Tanzania National Parks Authority (TANAPA), Tanzania Wildlife Management Authority (TAWA), and Ngorongoro Conservation Area Authority (NCAA)—each with their own stake in wildlife revenue, reflects the persistence of statism in the wildlife sector despite structural adjustments. Rather than diffuse state power, decentralization serves rather to consolidate it by enabling the state to efficiently "unlock" the value of mobile wildlife through a meshwork of formal institutions that ensure that revenues generated from wildlife tourism ultimately return to central coffers (Nelson et al. 2007). TANAPA deals with all wildlife-related issues that occur inside national park boundaries, including enforcing regulations and collecting tourism revenue, but their jurisdiction does not technically extend beyond the boundaries of national parks. In practice, this is not totally cut-and-dried, as TANAPA sometimes pushes for expanded park boundaries, enforces surrounding buffer zones, and provides services to adjacent communities like human-wildlife conflict mitigation. In a formal governance sense, however, and in terms of the conversion of wildlife to capital, TANAPA's jurisdiction over wildlife is circumscribed by park boundaries. As soon as wildlife leaves national parks, it falls under the administrative authority of TAWA.

TAWA extracts revenues from wildlife by issuing hunting permits in game reserves, GCAs, and open areas. Like national parks, game reserves are fortress areas that foreclose local resource uses and generate revenue from tourism, making them straightforward to administer from a governance perspective. GCAs, however, are more complicated. GCAs were devised by the state to allow TAWA (formerly the Wildlife Division) to extract wildlife revenue through the administration of trophy hunting blocks. Hypothetically, this meant from a statist perspective no competition from the community level. However, the initial GCA legislation from 1974 did not restrict human settlement, crop cultivation, and livestock grazing in the same fashion as game reserves and national parks. This made GCAs vulnerable to encroachment and alternative land uses. In many instances, this became a management issue as former colonial hunting blocks were increasingly cleared for agriculture, making them difficult to administer. In predominantly pastoral areas, the major source of tension was institutional layering (cf. Lesorogol 2022). Since the late 1970s, wildlife dispersal areas outside national parks that were designated as GCAs in accordance with the Wildlife Conservation Act often overlapped areas that could also be classified as villages based on the National Villages Land Act. This made it challenging for the central government to establish a sustainable model of extracting wildlife revenue from GCAs while also ensuring that rural communities did not lay their own competing claims to land. A key conflict in the context of GCAs that overlap village land is, What happens when trophy hunting companies have a concession within a GCA that has been issued by TAWA, but the village wants to use their land for other purposes, like livestock grazing or joint photographic tourism ventures with investors of their choosing? Both claims are legitimate to some extent, as GCA hunting blocks are administered by TAWA and communities have customary claims to village land (Nelson 2005; PINGO 2013).

In recent years, this issue has becoming increasingly pressing in Ngorongoro District, where Loliondo GCA has historically overlapped village land. From the perspectives of local Maasai, they view the land as part of their villages, having developed mutually beneficial, community-based photographic tourism partnerships in Ololosokwan and other villages (Kileli 2017). Under pressure from Ortello Business Corporation (OBC), an Emirati company, which has long held the trophy hunting lease for Loliondo GCA, the state has been inclined to side with the trophy hunters (Weldemichel 2020). This ultimately led to the burning of bomas in Ololosokwan by TANAPA rangers in 2017 and open conflict, which subsided in 2019 when OBC director Isack Mollel was arrested for employing foreign nationals without work permits. The conflict, however, resurfaced in full force following the passing of President Magufuli. In June 2022, violent confrontations broke out when state paramilitary forces demarcated the boundaries of Pololeti Game Reserve—formerly Loliondo Game Controlled Area—including 1,500 km^2 that overlapped fifteen villages in Loliondo Divison and historically provided crucial seasonal livestock grazing for pastoralists. Pololeti Game Reserve is currently managed in partnership with OBC as a private trophy hunting block for the global elite.

The state-led land grab follows a series of reforms to GCA policy in Tanzania to address the history of institutional layering and to consolidate state administrative authority. In 2000, the MNRT passed legislation prohibiting photographic tourism activities in hunting blocks and GCAs, followed by a stricter reform in 2009 prohibiting all local livelihood activities. This essentially made GCAs synonymous with game reserves in all but title. The firm-handed approach to steamroll over preexisting competing claims to these territories from the Land Act and Village Land Act signifies a clear move by the state to remap the institutional terrains of wildlife and land policy in a homogenous way. But like a "palimpsest," visible traces of former GCA policies still remain (Jackson 2021).

The exception to TAWA's control of wildlife revenues outside national parks is the NCA, which has its own administrative authority, NCAA, that is solely responsible for wildlife management inside the NCA, including enforcement of regulations and collection of tourism revenue. The NCA represents a progressive process of dispossession for the Maasai, who were evicted from the Serengeti Plains in the early 1950s, displaced from settling the Ngorongoro crater in 1975, and restricted from cultivating crops and building houses and schools, though the farming bans were lifted and reinstated several times in 1992, 2001, and 2009. Most recently, in 2019, NCAA proposed to evict ninety-three thousand Maasai pastoralists from the NCA entirely—yet another clear attempt by the state to eliminate competition for wildlife tourism revenue from local communities. Though its geographic jurisdiction is more limited than TANAPA and TAWA, NCAA generates by far the most revenue for the MNRT and thus has significant clout. Its political power has been increasing in Tanzania in recent years, and it is now responsible for administering the newly gazetted Pololeti Game Reserve, pointing to its expansion and diversification beyond just the NCA.

The distinctions between these bodies within the MNRT are meant in practical terms to allow the central government to consistently secure an unabated benefit stream from mobile wildlife resources through decentralization, though this is complicated in practice by occasional frictions between these arms over access to limited wildlife revenues—each institution has its own operating costs after all. Thus, despite neoliberalization, the socialist ideology forged by President Nyerere on the heels of a colonial legacy of centralization still permeates the fabric of resource governance in contemporary Tanzania. The state is reluctant to loosen its grip on wildlife resources by privatizing authority or devolving it to the level of communities. Rather, it strictly regulates the wildlife sector following a protectionist mindset that undermines the very prospect of neoliberal conservation. Of course, the state does forge strategic partnerships with private actors, provided that the arrangements function to secure a consistent revenue stream for government operations. But given a choice between supporting an equitable community-based conservation arrangement between a private investor and a village or establishing a new means of grabbing these lost revenues back into central coffers, the Tanzanian state seems consistently inclined toward the latter option. Such dynamics are difficult to explain under the umbrella of neoliberal conservation since the state remains deeply partial in every aspect of the wildlife sector. In such instances where revenues from wildlife begin to fall through its administrative cracks, the central government once again tightens the reins. And this, in a nutshell, is the backdrop of how and why Wildlife Management Areas (WMAs) were introduced in Tanzania in the late 1990s.

Recentralization and Wildlife Management Areas

State interest in monopolizing wildlife-related tourism revenue is particularly apparent when the government is faced with potential threats to central control in the form of private partnerships that emerge outside its administrative ecosystem. Direct investments in villages in the 1990s fell into this category since they did not neatly fit into the state's administrative silos of TANAPA, NCAA, or TAWA. Since the central government did not want to fully devolve ownership of wildlife to the level of communities, it devised a new institution informed in part by international trends in community-based conservation: WMAs. In technical terms, WMAs constitute multiple-use conservation areas on village lands that overlap wildlife habitat. WMAs are created by reclassifying lands shared by multiple "member" villages as a reserve through a process of land use zoning. The resultant management plan includes territorially defined restrictions on human activities designed to protect wildlife. Unlike other protected areas, however, management plans are meant to be designed and enforced by community members to ensure that livelihood activities that are important for human well-being are integrated in the conservation model.

The introduction of WMAs allowed the state to reconsolidate control over wildlife tourism occurring on village land by recentralizing tourism revenue collection. Tourists are charged entrance permits to visit WMAs, 35 percent of which is taken by TAWA, 5 percent of which goes to district government, and 60 percent of which is returned to the WMA. From this remaining 60 percent, half is allocated to cover the WMA's operational costs, and the other half is returned in equal distributions to member villages. Unlike the direct investment model, all revenue is centrally collected by TAWA before being dispersed back to WMAs and communities. WMAs thus undermine the village institution and strengthen state control over wildlife resources.

Though clearly an attempt to re-ascribe state authority over an increasingly neoliberalized wildlife sector, the initial motivations for the creation of WMAs were not entirely insidious, as there was also genuine interest in securing wildlife habitat outside national parks in the face of potential alternative land uses that might otherwise lead to ecological fragmentation. Furthermore, there was also pressure on the central government from the international community to devolve conservation governance in a participatory fashion to better represent the interests of communities. This was reflected in the revised Wildlife Policy of 1998, which stressed the importance of including local communities in the wildlife sector through the establishment of WMAs. The policy addressed the historical marginalization of communities in the wildlife sector and acknowledged their rights to utilize wildlife resources on village land. While the policy reinforced state control over all wildlife and land in Tanzania, it also formalized an institutional framework for local communities to extract revenue benefits from wildlife, at least in theoretical terms. This set the stage for land use plans to be drawn up at the local level, outlining parts of villages that could be set aside for conservation and the regulations that would be put in place to manage them.

WMAs, then, are microcosms of wider sectoral reform in that they are meant to address key social, economic, and ecological concerns that had arisen from the national park model of conservation (Nelson et al. 2007). They reflect the culmination of three intersecting interests: first, in securing the government's ability to extract revenue from wildlife on village land; second, in protecting wildlife outside protected areas from competing land uses; and third, in providing greater opportunity for communities to participate in the wildlife sector (Nelson et al. 2007). While some of these initial objectives of WMAs appear laudable, their political subtext is "thinly veiled," to borrow from Scott (1998). WMAs create a new platform for the state to reassert top-down control over village land. Following the Wildlife Policy of 1998, the central government put in place detailed procedures outlining how WMAs could be established. The government highlighted the need for multiple villages to unite in forming a community-based organization (CBO)—a clear signal that it was wary of empowering individual villages through the reform. CBOs function as a political body above the level of villages with the power to influence zoning plans on village land. While CBOs are responsible for creating the initial WMA land use plans, their proposals must ultimately be subjected to an environmental impact assessment carried out by technical experts and approved by the government, demonstrating how the area will be managed in a way that effectively preserves the local ecology. Only after this environmental impact assessment is complete can communities then apply directly to the Director of Wildlife to secure their rights to benefit economically from wildlife on village land (i.e., "wildlife user rights"). Thus, the WMA structure is based on a rigid legislative framework that was conceived from above by the central government as a means of taking back control of wildlife tourism on village land.

Rural communities in a contemporary context are invited to follow a predetermined set of procedures to establish WMAs. They must navigate various bureaucratic hurdles to do so, without any real guarantees that doing so will secure their resource rights rather than undermine them. Much of the conflict that has been documented in the literature on WMAs in Tanzania stems from uncertainty around what the exact implications of WMAs are for communities. As Wright (2019) documented in his ethnography of Enduimet WMA, the local Maasai community in Longido struggled for a long time to decipher whether the WMA institution represented a "friend or foe," in that they wondered whether it would secure pastoral tenure and protect their livelihoods or displace them (chapter 3). Maasai leaders were initially very wary of WMAs because the idea to form them, and the governance structures for operating them, originated from the state. Within WMAs, community members are not able to articulate the fundamental conditions of their wildlife user rights without seeking permission from central government authorities. In the early years following the wildlife policy reforms, pastoral land rights NGOs like UCRT were concerned that WMAs might simply allow the central government to establish protected areas in villages that served to further alienate pastoral land and siphon tourism revenue back to the state.

Rather than form a WMA, Maasai leaders in Loliondo sought to instead secure CCROs, procure direct investments in villages, and actively manage wildlife on village land for their own benefit based on their own bottom-up governance and management plans (Gardner 2016). They worried that like other strategies employed by the Tanzanian government throughout history, WMAs could be used in ways that were antithetical to the interests of pastoral communities. Local activists cautioned that WMAs were not, in and of themselves, community-based models of conservation, but rather a particular form of wildlife conservation conceptualized by the central government to forward its own interest in maintaining state control of resources during a period of neoliberalization (see Wright 2017). Maasai leaders in Loliondo reasoned that the primary purpose of WMAs was not to protect community livelihoods, but to enable the central government to extract revenue from wildlife that moved outside the jurisdictions of TANAPA, NCAA, and TAWA onto village land. This background is deeply important and validates much of the concern that local communities have with the concept of WMAs. In mid-2020, I attended a workshop in Mto wa Mbu ("river of mosquitoes") town led by UCRT that brought together Maasai traditional leaders from across northern Tanzania to discuss strategies for revitalizing customary rangeland governance and management institutions. One of the key concerns raised by the leaders in attendance was that once a WMA is formalized, there is little recourse for villages to back out if it becomes evident that the WMA does not represent local livelihood concerns. Maasai communities must thus be as certain as they can be that WMAs are truly in their interests.

Confident that villages and CCROs would provide greater tenure security than WMAs, Maasai communities in Loliondo notably refused the establishment of a WMA despite pressure from the state and international NGOs like World Wildlife Fund and the African Wildlife Foundation (Gardner 2016). The wildlife policy, however, made clear that direct investments in villages and WMAs were mutually exclusive. By bringing WMAs into being, the state foreclosed direct investments in villages and required that all community-based wildlife tourism on village land occur within the context of a WMA. By refusing a WMA, the Maasai in Loliondo signaled to the state that they would not subscribe to the new model of community-based conservation and would continue to struggle for sovereignty through the village institution. The state ultimately responded by reforming its GCA laws in 2009 and in 2022 began converting GCAs into game reserves that afforded local communities *no* wildlife user rights whatsoever. In overlapping areas, like Loliondo, the state grabbed the land back from communities through top-down force. This was the political backdrop of the violent conflicts that broke out in Loliondo in 2022 when the state formally gazetted Pololeti Game Reserve in the former Loliondo GCA.

WMAs thus represent an existential dilemma for the Maasai of Tanzania. Communities can reject them and work to secure village titles instead. Though direct tourism investments in villages are prohibited, resistance is always a possibility, in Loliondo and elsewhere. Relying exclusively on the village institution runs the risk of communities being evicted from wildlife-rich areas in favor of game reserves or other centrally managed protected areas since villages alone do not afford communities wildlife user rights. Alternatively, communities can establish WMAs and risk undermining the power of villages for governing pastoral territories, but perhaps gaining in the process a formal institution for managing land *and* wildlife that is backed by the state.

The current anthropological literature on WMAs in Tanzania's Maasailand now includes three notable cases of Maasai communities interpreting these policy reforms in distinctive ways and determining whether or not to implement WMAs: Loliondo, Longido, and Lolkisale. While Maasai in Loliondo rejected the WMA and chose to rally around the village institution, Kisongo Maasai of Longido and Lolkisale followed a different path, choosing not only to implement WMAs on village land, but to wield them strategically as formal institutions for governing and managing pastoral territory. The remaining chapters of this book unfold the history and cultural politics of Randilen WMA in Lolkisale against the backdrop of this ongoing struggle for sovereignty between Maasai communities and the state, and the resonating question of whether WMAs help the Maasai secure tenure and benefit from wildlife on community land or further dispossess and marginalize them.

CHAPTER TWO

The Lolkisale Land Squeeze

Prior to colonialism, the Kisongo were the dominant group in Lolkisale, having pushed out, or assimilated, the Mbugwe, Barabaig, and Akie who inhabited the area prior to their arrival. While the area was used for seasonal grazing as part of the pastoral mode of production, it had high densities of tsetse flies, which limited human settlement patterns and shaped local herding practices (Arlin 2011). For the most part, the Kisongo used the area for dry season grazing, particularly in times of drought. Up until the 1950s, the area was sparsely populated, and the Kisongo kept it aside for reserve pasture to be used in times of dire need.

A series of land policies during the mid-colonial period began to constrain pastoralists in Lolkisale. Tarangire and Lake Manyara reserves were gazetted in 1957, and Tarangire was upscaled to a national park in 1970 (Davis 2011). These changes dispossessed herders from key sources of water and pasture and created an arbitrary political division between "pristine wilderness" spaces and human society (Igoe 2004, 2017; Igoe and Brockington 1999; Neumann 1998). The land squeeze was exacerbated by the reallocation of productive rangelands near Lolkisale mountain for settler farms in the 1950s. Domestic laborers of mixed ethnic backgrounds migrated to Lolkisale to work on these farms, eventually leading to the development of the town-like centers in Lolkisale-proper and Nafco that are seen today.

Following independence, sizeable areas in Lolkisale were set aside by the state for large-scale agricultural production and leased to private companies (Wøien and Lama 1999). In 1971, commercial bean and flower seed production commenced in Lolkisale, expanding to over 6,500 hectares by 1976 (Borner 1985; Gardner 2007). Commercial seed production for export resulted in large-scale land cover change in the 1970s–1980s. Under the title of the Rift Valley Seed Company Ltd., Hermanus Steyn was allocated the entire

A Kisongo Maasai man guides the anthropologist to key sites in Lemooti village, with Lolkisale mountain in the background. Photo by author in 2019.

A view from partway up the mountain shows farmlands enclosing Lolkisale village. Photo by author in 2019.

area bordering the northeastern boundary of Tarangire NP, a parcel of around 381,000 acres that encompassed the current-day villages of Makuyuni, Naitolia, Nafco, and Lolkisale (Homewood 1995; Wøien and Lama 1999). Though Steyn intended to intensively develop the land for agriculture and ranching, he had only cleared small portions of the area for bean farms by the early 1980s when the government abruptly (and unceremoniously) canceled his lease. Steyn appealed the government's decision, and the dispute remained unresolved through the 1980s and escalated to national controversy in 1994 when the Parliament of Tanzania ultimately voted against him retaking possession of the land (Wøien and Lama 1999). Though Steyn was partially compensated for the loss of land, his grievance was still unsettled at the time of writing.

While commercial agricultural companies like Steyn's Rift Valley Seed Ltd. were the major players involved, smallholder farmers were also implicated through growing contracts. The Sluis Brothers Ltd., for instance, provided small producers with "stock seeds" to plant on their farms (Nshala et al. 1998, 71). In exchange, Sluis Brothers held the right to purchase back the harvests. One of my interviewees from the Esimangore sub-village of Makuyuni reported participating in a growing contract of a similar nature through the 1980s. Including Steyn's farms, which were nationalized in the 1980s, bean seed plantations covered 25 to 55 percent of contemporary Makuyuni, and the entirety of contemporary Nafco (Borner 1985).

Further complicating land issues, villagization in the 1970s created distinct administrative units for allocating land at the local level. Though Operation *imparnati* (chapter 1) had limited effects on Kisongo settlement patterns in the Lolkisale area, villagization further spurred in-migration of Arusha from Meru into the area, allowing them to collectively increase their total land under cultivation. Arusha leveraged social (intermarriage), cultural (shared language), and political (leadership positions on village councils) influence to gain access to the area (Kuney 1994). The Arusha were also ideally positioned to capitalize on the cancelation of Steyn's lease. Following the government's decision, Steyn's property holdings were initially nationalized, and the National Agricultural and Food Corporation (NAFCO)—from which Nafco village takes its name—and the Monduli Development Corporation (MODECO) took over management of the farms for a period before turning much of the land over to local villages. Farms spanning Nafco, Makuyuni Juu, and Naitolia were subsequently designated as village land. The Arusha accelerated their patterns of in-migration during this period first to work on the MODECO farms and later to stake individual land claims informally through cultivation

(Kuney, personal communication, 2019). Many did so without paperwork or formal processes of allocation through villages or other channels. Arusha encroachment continued into the 1980s while the government carried out land surveys across the Maasai Steppe to enable village registration and titling (Wøien and Lama 1999). Rangeland fragmentation became more entrenched in the 1980s as national development policy turned in favor of economic liberalization. With government encouragement, farms were cleared north of Lolkisale, near the Engorika Hills, and by the 1990s, large areas in Mswakini had also been alienated by commercial farms. In the 1980s–1990s, the government increasingly supported foreign investment and mechanization of agriculture through lease contracts with private companies. Charcoal production also became widespread, with rising demand in urban centers driving deforestation by small-scale producers in the areas adjacent to Tarangire NP.

Managing Dispersing Wildlife

Land use change in the areas east of Tarangire NP posed a great conservation challenge for the government in the 1980s. While the Maasai Steppe at large spanned an area of approximately 40,000 km², only 2,600 km² was located inside Tarangire NP. As mentioned in the introduction, wildlife generally disperses outside the park in the wet season before returning to its permanent water sources in the dry season, including the Tarangire River and key wetland swamps like Ngusero and Lormakau in the south, Silale in the east, and Gursi in the west (Bond et al. 2022; Kiffner et al. 2022). In the wet season, the Simanjiro plains to the east of the park serve as crucial calving grounds for zebra and wildebeest, and Lolkisale and Makuyuni are important dispersal areas for the largest subpopulation of elephants in northern Tanzania (Foley and Foley 2022). Evaluating the wildlife movement routes outside Tarangire NP in the early 1980s, Borner (1985) painted a grave picture of the impacts of commercial agriculture on the ecological connectivity of the Maasai Steppe. In his words, "the north-eastern migration [through Lolkisale] is seriously hampered by seed-bean farming north of Lolkisale Mountain and will be blocked completely in the near future" (Borner 1985, 92). Thus, from a conservation standpoint, commercial agriculture posed a crisis for Tarangire's wildlife.

While the land squeeze had been caused for the most part by large-scale alienation of settler farms in the colonial period, and commercial seed production in the 1970s–1980s, the government focused its attention on the clearing of wildlife habitat by smallholder cultivators. While this was indeed a growing concern, it was perhaps not the root of the problem. Nonetheless, Tanzania National Parks Authority (TANAPA) went to great lengths in the 1980s to try to convince Maasai communities living east of the park not to farm. In Lolkisale and other villages adjacent to Tarangire NP, TANAPA authorities sought in the 1980s and 1990s to offset the costs of dispersing wildlife for Maasai communities and help villages harness some of the benefits. TANAPA implemented the Community Conservation Service to provide support to communities adjacent to Tarangire NP. TANAPA reasoned that addressing the costs of wildlife conservation might incentivize people living near the park boundary to stop clearing wildlife habitat for crop production. Local community members refer to this as "being a good neighbor" (*kuwa ujirani mwema*), a sentiment that shone through in my interviews about Tarangire NP in villages adjacent to the park (Nshala et al. 1998, 24). TANAPA seemed to grasp the fact that strong-handed approaches to enforcing wildlife laws were making rural communities resentful toward conservation (Nshala et al. 1998). Between 2000 and 2005 in Simanjiro District, TANAPA spent just over $150,000 on community projects aimed at improving local sentiment toward wildlife conservation (Nelson et al. 2010). Local Maasai were particularly appreciative of TANAPA's efforts to address livestock depredation by large carnivores, provide water for livestock, and help with security issues in the context of potential cattle rustling (Kipuri and Nangoro 1996).

At the same time, local pastoralists also had significant concerns with TANAPA. Through the 1980s and 1990s, the community of Lolkisale regularly complained that they were being denied grazing access to open areas by TANAPA staff and that communal pastures were being alienated by commercial farms (Nshala et al. 1998). While TANAPA staff could not offer much in the way of support for the farming claims, Lolkisale villagers felt that it was government officials who had secured the farms at the expense of village communities and held that TANAPA should take responsibility on behalf of the government since it was an arm of the state (Nshala et al. 1998). Villagers lamented "empty promises" of *maendeleo* ("development") by the government that never came to be, and TANAPA served as the visible scapegoat for these feelings of resentment (Nshala et al. 1998, 25).

Perhaps the most significant grievance that my interviewees raised about Tarangire NP was that they were unhappy with the way that TANAPA consistently expanded its boundaries to grab more pastoral land even while claiming to help them (Igoe 2017; Igoe 2022; Sachedina 2008). Lolkisale community members were particularly concerned about these transgressions in the late 1990s, and during my interviews in Mswakini Juu, interlocutors mentioned that some of these boundary issues were still ongoing in 2020. When I interviewed the director of Tarangire NP in early 2020, he explained that the main issue from his perspective was that national parks are supposed to have a 2 km buffer zone around their boundaries where human settlement and cultivation are prohibited, but neighboring villages had expanded right up to the boundary of the park. Buffer zones created ambiguity as the communities felt they were part of village land, while TANAPA authorities were trying to enforce them as an extension of park boundaries. Technically, however, TANAPA could not enforce this legislation or provide services to communities in areas where it did not have legal jurisdiction. TANAPA staff could not formally work outside the park, so their enforcement of conservation regulations in the buffer zones was uneven and inconsistent.

As a consequence of TANAPA's lack of jurisdiction outside Tarangire NP, community conservation services had to be carried out through village-level administrative structures, which were constrained by limited capacity and start-up capital. Neighboring communities struggled to realize benefits from wildlife on village land. Conservation remained something that they were told to support by the government, but with little in the way of incentives. Pastoralists and smallholder farmers alike associated wildlife with livelihood costs and top-down land grabbing, but not with its potential benefits. Several studies from this period indicate that wildlife in Lolkisale village was having significant negative impacts on people's livelihoods. A study from 1999 indicated that 84 percent of people surveyed in Lolkisale had negative attitudes toward wildlife (Wøien and Lama 1999). Another study based on field research in the early 2000s revealed widespread discontent with the costs wildlife were having on household economies and the lack of benefits they were generating (Rija 2009). While these sentiments are troubling for human rights reasons, TANAPA and the conservation community were becoming increasingly concerned about what the long-term impacts of human-wildlife conflict would be for the Tarangire ecosystem as a whole. Disenchantment with wildlife at the community level was spurring pastoralists east of the park to clear land for agriculture in the 1980s in an attempt to actively steer wildlife away from villages (McCabe, personal communication, 2018). It was also a means of laying claim to land in the face of land grabbing and Arusha encroachment, and helped community members stake out material boundaries between villages and wildlife areas. Attempts to block dispersing wildlife through defensive cultivation practices were highly practical: Why would herders want wildlife on village land if they were unable to harness the benefits of these resources while being forced to bear the costs? Cultivation, then, became a tool to insulate community land from intruding wildlife, posing a significant concern to conservationists who worried that land outside Tarangire NP was becoming increasingly fragmented by expanding human settlements and agriculture. Reflecting on these trends, Borner (1985, 91) famously wrote about the "increasing isolation of Tarangire National Park," and conservationists began to realize that the existing institutional arrangements for governance and management were inadequate for dealing with the complex challenge posed by a dynamic ecosystem that extended beyond park boundaries.

While Borner's (1985) article was influential in inspiring the emerging discourse in Tanzania on wildlife corridors, close reading reveals a diplomatic tone that is appreciative of the potential for Maasai pastoralists and wildlife to coexist in the areas east of Tarangire NP. As he noted, "although the Maasai in Simanjiro suffer from the occasional attack on their livestock by lions or from elephants damaging their wells, they experience no forage conflict and they still favor the coexistence of game and livestock. There seems to be no serious objection to a dual use of the Simanjiro plains and the Lolkisale GCA for livestock and wildlife" (Borner 1985, 95). In line with these sentiments, a proposal was put forth in the early 1980s to establish a multiple land use authority in the Lolkisale area that was similar in management structure to the Ngorongoro Conservation Area (Bluwstein 2022; Borner 1982). At the time, projects were underway to map out land uses and biodiversity of the areas adjacent to Tarangire NP, including one called the Tarangire Conservation Project, led by an Italian nongovernmental organization (NGO) (Bluwstein 2018a). The multiple land use proposal drew from these insights and recommended establishing a 6,000 km² reserve encompassing Lolkisale and surrounding areas to protect wildlife moving outside Tarangire NP. Though TANAPA supported the proposal, it was ultimately rejected by the Ministry of Livestock. A similar proposition was tabled by Prins (1987), outlining migratory routes of large mammals outside the park and specifying land uses that were compatible with the preservation of 35,000 km² of wildlife habitat. The plan included bans on cultivation and reduced livestock numbers, among other recommendations. In the end, neither proposal was put into practice because they required land use planning efforts that were difficult to fashion into the vertical silos of the government's central ministries (Wøien and Lama 1999).

For the sake of pastoralists in Lolkisale, it was likely a blessing that the plans were never carried out, as they were designed by wildlife conservationists without input from local community members. The proposed restrictions would have likely had significant impacts on pastoral and agricultural livelihoods, further constraining the local economy. At the same time, the idea of a multiple land use plan in the Lolkisale area was not necessarily antithetical to the interests of the Maasai community. As described to me during interviews in Lolkisale, many of the Kisongo elders greatly resented the encroachment of Arusha cultivators onto their territory and the blockage of livestock corridors in the area, which affected pastoralists and wildlife alike. Harking back to days of the Masai Reserve, the Kisongo were very concerned about maintaining the livestock corridor connecting northern and southern Maasailand, which was becoming squeezed by commercial and smallholder agriculture on either side (Igoe, personal communication, 2019). The last remaining corridor connecting Simanjiro up to the north was via Lolkisale. As outside interests in pastoral lands from farmers, development organizations, and the central government continued to increase, the Kisongo lamented the lack of formal structures to defend their pastoral land from being grabbed and converted to other uses. The interests of the Kisongo in Lolkisale in protecting pastoral livelihoods thus reflected clear common ground with the wildlife conservationists who were keen on preventing further agricultural enclosure of Tarangire NP. For various reasons, however, the government failed to nurture these shared objectives, leaving a promising and underexploited niche for private companies to explore.

A New Tourism Frontier

While the ecosystem was becoming increasingly fragmented in the unprotected areas outside Tarangire NP, tourism inside the park was growing rapidly. Between 1992 and 2006, Tarangire NP and Lake Manyara NP received nearly two million tourists combined, generating revenue that rose almost eightfold over the period (Sachedina and Trench 2009). Tarangire NP specifically saw its number of yearly tourists quadruple from 1989 to 2002, with total revenues rising 3,650 percent over the same period (Nelson 2004; Rodgers et al. 2003). From the perspectives of tourists and tour operators, however, the park experience was limited. Game drives took place on fixed circuits and in some ways paled in comparison to the diverse tourism opportunities that existed on community lands adjacent to the park (Nelson 2003, 2004). Unrestricted by park regulations, tour operators outside the park could offer cultural experiences, horseback riding, walking safaris, night drives, fly camping, and even off-road game drives under the supervision of local Kisongo guides. Given the vast and dynamic nature of the Tarangire ecosystem, many villages hosted as much, and sometimes more, wildlife on a seasonal basis than the park itself (Goldman 2018). They also offered the added appeal of fewer tourists and infrastructure, less "congested" game viewing, and a more "exclusive" experience (Nelson 2004, 6).

These selling points were strong enough for several investors to approach villages east of Tarangire NP with proposals about establishing private tourism arrangements on village land (chapter 2). In Lolkisale, there were three main factors that led to the emergence of direct partnerships between investors and the village council. First, there was shared interest among the Kisongo and tour operators in preventing agricultural conversion. For the Kisongo, concessions protected pasture for livestock, and for the tour operators, they preserved habitat for wildlife. Second, both parties had interest in capitalizing on the abundant wildlife that was dispersing onto village land. For community members, this would offset some of the costs of wildlife for household economies and community livelihoods; for investors, community-based ecotourism offered a potentially lucrative opportunity to diversify the tourist experience away from game drives in the park. And third, TANAPA was unable to capture this emergent market. These three factors set the stage for an opportunistic investor to forge new ground in Lolkisale village in the late 1980s and early 1990s by developing a private concession on village land in collaboration with Lolkisale Village Council (LVC). The area, which sought to address the dual aims of protecting wildlife habitat and pastures outside the park, later became called the Lolkisale Conservation Area (LCA).

The original impetus behind the LCA traces back to a partnership between an Australian investor, his brother, and Lolkisale village in the late 1980s and early 1990s. The investor came to establish the East African Safari & Touring Company (EASTCO), a family-run business based out of Arusha that today (2022) operates and manages lodges and safari camps in Tanzania for wholesalers overseas (LBCSP 2003, 9). EASTCO's selling points are its "off the beaten track" "wilderness" experiences in the Tarangire ecosystem. These often involve walking safaris and fly camping on village land. When the investor first approached Lolkisale village, his company was just a one-man, one-car operation before blossoming in the late 1990s. In 1997, EASTCO launched its online website and has since developed into a major tour operator throughout East Africa, scaling up from two initial lodges in the Tarangire ecosystem to over fifteen across the wider region.

In the late 1980s, the Australian brothers became increasingly concerned that the dispersal routes for wildlife moving outside Tarangire NP were becoming blocked by commercial farms. Coupled with this, they recognized the potential for developing an ecotourism enterprise outside the park that offered something creative and new. While visiting a friend's farm in Lolkisale in 1993, they developed the idea to establish a joint conservation venture with the local Kisongo community and set their sights on Boundary Hill, a stunning viewpoint in Lolkisale village overlooking Tarangire NP. They negotiated with LVC for about 2,000 acres of land around Boundary Hill where they wanted to establish a luxury tourist lodge. They also established a small four-bed bush camp in Lolkisale called *Sidai* Camp ("things are good" in Maa). The brothers planned to carry out walking safaris, fly camping, and night game drives in Lolkisale as part of the agreement and sought permission from the village council to build the lodge on the hill and secure exclusive tourism rights to the area. Part of the agreement involved setting aside a concession area that would not be farmed by the local community. In exchange, Lolkisale was to receive a share of the tourism revenues. A key part of this proposal was that it was legally structured as a joint venture between LVC and the Tarangire Conservation Co. Ltd. (TCCL), a subsidiary company wholly owned and operated by EASTCO (LBCSP 2003, 9). Lolkisale village and TCCL were registered as equal 50:50 shareholders in the new company, which they subsequently named Boundary Hill Lodge Co. Ltd. This form of community-based conservation contract was groundbreaking at the time, as it meant that profits were to be split evenly between TCCL and Lolkisale village. The joint venture was formalized in 1995, and the community of Lolkisale began putting local restrictions on farming in the designated concession area. At the time, Lolkisale had not yet subdivided and included the sub-villages of Lemooti, Lengoolwa, Nafco, and Oldonyo ("mountain" in Maa). It was also predominantly a Kisongo village with a population of around six thousand people, though the Arusha were established in some of Lolkisale's sub-villages at that time as well (LBCSP 2003). In 1998, LVC signed a ten-year lease agreement with Boundary Hill Lodge Ltd. that entitled the joint venture to exclusive photographic tourism rights over the area, including the rights to tented camping and walking safaris (LBCSP 2003).

While Lolkisale village agreed to the terms set forth by TCCL, the investor lacked the start-up capital necessary to actually build the lodge. To raise funds, Boundary Hill Lodge Co. Ltd. applied for a Small and Medium Enterprises loan from the International Finance Corporation, a division of the World Bank Group. The investor framed the proposal as a community-based conservation venture that would benefit the environment by preventing biodiversity loss while simultaneously contributing to sustainable development at the village level. They were awarded $200,000 in support of their plans to build the Boundary Hill Lodge (LBCSP 2003). The community of Lolkisale was supportive of this plan, hoping to benefit from the revenue stream of the subsequent lodge. The limited funds, however, put a drag on lodge construction, which progressed slowly into the early 2000s.

The Australian investor's interest in the community lands northeast of Tarangire NP were not limited to Lolkisale village. In 1996, he entered into another agreement with the villages of Makuyuni and Mswakini Juu (prior to subdividing into Naitolia) to establish the Naitolia Concession, an exclusive tourism zone adjacent to Lolkisale (PINGO 2013). Based on the agreement, the two villages set aside 13,590 acres for photographic tourism and walking safaris. Located within the concession were the Lemiyon plains and the Naitolia floodplains, which were ecologically significant for wildlife. The concession demarcated an exclusive-use zone for a small eight-bed bush camp called Naitolia Camp, though the investor held the rights to establish another tented camp in the area if desired. Two potential camp sites were selected by the investor through consultation with the communities to ensure that their grazing and farming rights were maintained, but only one was ultimately developed. An area of 2,000 acres was demarcated around the camp sites where grazing was prohibited. The investor agreed to pay a flat rate of $1,000 per year for exclusive use of the area, split evenly between Makuyuni and Mswakini. The initial contract was for five years, with the possibility of renewal if the villages and investor agreed.

Eager to get the luxury lodge at Boundary Hill up and running, and perhaps recognizing that $200,000 was a fairly modest amount for an initial investment in the area, the investor applied for additional funding from the Global Environment Facility (GEF) in 2003. He was subsequently awarded $35,000 in technical assistance funding and $450,000 via a Multiple Stakeholder Partnership (MSP) platform loan. As stipulated in the proposal, the funding was meant to establish and implement an Integrated Conservation Management Plan for a scaled-up LCA. While the technical assistance funding was meant to finance the design of a management plan for 40,000–60,000 acres around Boundary Hill Lodge, the MSP loan was provided to implement a large-scale integrated conservation management plan across 145,000 acres adjacent to Tarangire NP. The loan was intended to finance a multiple stakeholder approach to streamlining tenure policies in the area and establish a sustainable revenue stream that was to be equitably shared between investor and communities. From the perspective of GEF, the proposal was appealing because the funds would be used to support the protection of a key wildlife dispersal area and contribute to institutional harmonization of conservation governance and management outside Tarangire NP.

Lolkisale village did not take the prospect of an expanded conservation area lightly. Prior to the submission of the MSP loan proposal in 2003, Lolkisale's leaders systematically weighed the benefits and risks of the LCA in 2001–2002. It was democratically decided, through the governance structures of the village council and with the support of the assembly, that the community of Lolkisale would support the LCA. The village council held numerous assembly meetings to discuss the concept of a community-based conservation area and the benefits of establishing the LCA (King 2009). A secretary diligently documented these meetings with minutes to ensure due process was upheld. By June 2001, a detailed management plan for the conservation area had been drafted that clarified the potential economic benefits that the community of Lolkisale would accrue and the conservation and environmental outcomes that would result for the ecosystem at large. On December 20, 2001, the Ward Development Council held a key meeting that formalized the overall LCA area, including the zoning scheme outlined in the tentative management plan. It was considered a significant occasion for all involved stakeholders, as it represented an exciting new community-supported direction for conservation outside national parks (King 2009). By February of 2002, the Monduli District Council had formally given the go-ahead to commence conservation and tourism activities on the ground in the LCA based on the management plans that had been designed collaboratively by the community, the investor, government stakeholders, and international donors.

One of the key aspects of the expanded LCA was the creation of a large buffer zone adjacent to the initial conservation area called the Lolkisale Livestock and Wildlife Zone (LLWZ), encompassing 99,000 acres. The Kisongo saw it as a means of defending the land from encroaching cultivators, and the investor saw it as a way to secure wildlife habitat in a key dispersal area. It was managed by local pastoralists through the Integrated Conservation Management Plan, which essentially meant maintaining it as common pasture. Importantly, from the perspectives of local Kisongo, the LLWZ covered the last remaining part of the livestock corridor connecting northern and southern Maasailand via Lolkisale. Further to the LLWZ, the Makuyuni Elephant Dispersal Area (MEDA) was added to the LCA management plan in 2001. MEDA overlapped a communal grazing area in Makuyuni village called *ndoroboni* and was added based on the recommendation of the Tarangire Elephant Project, a long-term ecological monitoring initiative started in 1993 with financial support from the Wildlife Conservation Society (Bluwstein 2018a). Realizing its importance for elephant population persistence, the Elephant Project lobbied for a key elephant habitat area in Makuyuni to be included in the

LCA. The area was identified as an important wet season dispersal area for Tarangire's northern subpopulation of elephants. Makuyuni agreed to set aside 11,000 acres for MEDA. In exchange, Makuyuni village council was interested in partnering with a prospective investor, as the village had already been receiving some modest revenue from the Naitolia Concession.

Together, the LCA, LLWZ, MEDA, and Naitolia Concession formed the Tarangire Conservation Area, an area that would be managed holistically as part of the integrated management plan funded by GEF (King 2009). By 2000, the Tarangire Conservation Area spanned just over 144,000 acres, including four distinct zones that had been demarcated by the villages and the Australian investor for community-based conservation and ecotourism: LCA (40,500 acres), LLWZ (99,000 acres), MEDA (11,000 acres), and the Naitolia Concession (13,590 acres). Collectively, the four areas protected key ecological features and wildlife habitat, including the northern watershed for the Tarangire River and Gosuewa swamp (LCA), the Lemiyon Plains and Naitolia floodplains (Naitolia Concession), and crucial dispersal areas for elephants (MEDA), wildebeest, and zebra (LLWZ).

While the LCA was an exciting prospect on paper, in practice, Boundary Hill Lodge struggled to get off its feet. Though EASTCO's website stated that Boundary Hill Lodge was ready for guests by 2002, it was actually still in the throes of construction. Building the lodge was challenging, due in part to Boundary Hill's remote location and the investor's limited start-up funds. Things turned for the worse in 2005 when the lodge burned down, delaying construction further. While the origins of the fire are unconfirmed, the Australian investor alleged that the lodge was burnt down by hunters who resented the photographic concession around the lodge. According to an account relayed to me by a district government official who was knowledgeable of the situation as it unfolded, however, the investor's brother may have burned down the lodge ahead of a visit from World Bank representatives, knowing that the investor had only used a portion of the money to build the lodge and had spent the rest on other ventures overseas. I am unable to substantiate whether either allegation holds merit. Irrespective of the fire's cause, Lolkisale village was disappointed in their Boundary Hill joint venture because it was not offsetting the opportunity costs of keeping their land aside for the concession. People in Lolkisale began calling Boundary Hill "*tasa*," which translates roughly as "unproductive" (PINGO 2013:4). While Lolkisale village and TCCL were technically 50-percent stakeholders in the project, community members felt that the investor had leveraged the venture to secure the loan only to use it unsystematically to build Boundary Hill Lodge and establish Naitolia Camp without adequately compensating the community in return. Beyond what I have been told by community members and district officials, I have not personally seen evidence that the investor used the money in ways other than what he had initially proposed to do: build a lodge and implement a management plan for the Tarangire Conservation Area. What is perhaps likely, however, is that community resentment stemmed from a lack of effective communication and frustrated expectations of tourism revenues that took a long time to materialize. Lolkisale village was keen on receiving a head fee from visiting tourists to contribute to community services and infrastructure development projects, but these payments could not be made until the lodge was built and ready to host visitors.

While the investor may well have used the money appropriately to accomplish the plan he originally set out to, from the perspective of community members in Lolkisale, there was a lack of transparency in how the money was used. This led people to distrust the Boundary Hill Lodge venture, and a feeling began to take root across the community that Lolkisale village had been swindled. Prospective tourists also expressed dismay at the state of Boundary Hill Lodge, as evidenced by a heated discussion on an online travel forum from 2004 to 2006. As one tourist wrote on February 4, 2006, after booking a trip to the lodge via the EASTCO website,

> Their lodges/camps are a mystery to myself and everyone I talk to (especially when compared to the other camps in the vicinity)....I merely think S***n should update his web sites and provide factual information about the status of his camps/lodges in Naitolia, Boundary Hill and Sidai. This is not too much to ask! For example the website currently states that Boundary Hill will open in 2002! This is misleading as the lodge has never been completely constructed! Why is everything so secretive? Myself and many on this board will strongly support his efforts but he needs to do his part as well! The local villages (most certainly with high hopes) deserve better. (Fodor's 2006, 15)

Distrust reached the point of widespread worry when Lolkisale village got wind of the fact that the investor had defaulted on his loan from the World Bank, potentially leaving the community on the hook to pay back the difference (~$700,000) as 50-percent partners in the initial proposal. This led to considerable fear within Lolkisale that community land would be taken away by the World Bank Group as compensation to pay back the loan. Despite a period of worry, the World Bank was ultimately made aware of the community's concerns and decided to turn over the loan to Lolkisale village. Thus, rather than pay back the loan to the World Bank, the investor became legally obligated to pay it back to Lolkisale village, though at the time of my initial fieldwork in 2019–2020, Lolkisale village had still not received the money. The village sued the investor over money owed, and the case was finally closed in 2023 with the court ruling in the community's favor. As of 2024, however, the community had still not been paid.

In the early 2000s, frictions also emerged at Naitolia Camp. As a bush camp, Naitolia Camp did not require as much start-up capital and was quick to get off the ground. However, there was no agreement in place for sharing tourism revenues with the villages, as the concession had been negotiated based on a flat rate (PINGO 2013). Concession fees were used to finance community development projects, which villagers appreciated. These included the installation of a pump in Makuyuni River to supply Naitolia and Makuyuni with water, the construction of a health dispensary in Naitolia, and the purchase of a community tractor. But since the concession rate was low and the villages had subdivided into Makuyuni, Naitolia, and Mswakini Juu, the funds were minimal from the perspective of villagers. Village councils wanted a share of the revenue streams, but in all fairness, the camp only hosted eight guests and did not generate much profit at the best of times. As part of the revised Tarangire Conservation Area Management Plan of 2009, additional tourist camps were proposed in the Naitolia Concession and MEDA to diversify the revenue streams for these villages. Makuyuni wanted to partner with Kikoti Lodge, which was already operating in Loiborsoit near the Arusha-Manyara regional border, but this never came to fruition. Ultimately, the Naitolia bush camp struggled to sustain itself and was abandoned.

Although the Australian investor had not delivered on the promise of a "win-win" joint venture, he indirectly contributed to one of the most successful investor-village partnerships in all of Tanzania. While Lolkisale villagers patiently waited for Boundary Hill Lodge to be constructed in the late 1990s, the investor's brother raised the idea of establishing another lodge within the LCA with the permission of LVC. The lodge was proposed as a means of providing more immediate revenue to the community from a company that had the capital on hand to build it. This would help reduce the opportunity costs of not clearing land for agriculture in the concession zone while Boundary Hill Lodge was being built—a "stop-gap" so to speak. After negotiating the terms of this arrangement with LVC, Boundary Hill Lodge Ltd. sold a short-term lease concession of 100 acres to Halcyon Tanzania Ltd. to build Treetops Lodge (LBCSP 2003). Halcyon Tanzania Ltd. partnered with Boundary Hill Lodge Ltd. to build the lodge in 1999, and the original investor's brother negotiated the terms of the benefit-sharing structure with Lolkisale. Since Treetops was meant to be a short-term solution to benefit the community while Boundary Hill Lodge got on its feet, it was determined that a $15 head fee for each overnight guest would be paid by Treetrops Lodge directly to Lolkisale village. Locally, these payments are referred to as "bed night fees." Treetops's model of direct payments to the village for each visitor differed from the benefit-sharing structure institutionalized by Boundary Hill Lodge Ltd., which collected all bed night fees with the plan of reinvesting them in LCA management. While Treetops generally hosted guests for a few nights at a time, it was hoped that Boundary Hill Lodge would later fill out the market for longer stays. The uniqueness of Treetops, named for its elevated perch within the canopy of large baobab trees overlooking the savanna, appealed to adventurous safari-goers, and the forty-bed lodge attracted a large number of guests in the early 2000s. Much to the appreciation of community members, Lolkisale accrued approximately $50,000 annually directly to its Community Development Fund from its partnership with Treetops, a figure that would rise to almost $80,000 by 2008. By that point, Treetops had been sold to Elewana Afrika Ltd., which maintained the lodge's positive relationship with Lolkisale village and continued to share the pre-negotiated bed night fees with the community.

Revenue from Treetops had a major impact on people's well-being in Lolkisale village. A key reason for this was good governance at the local level. Since the village was first established following villagization, the Lolkisale assembly has always elected the same chair. He is a humble man, whom the community holds in high esteem because of his steadfast dedication to ensuring that democracy prevails. My interlocutors in Lolkisale describe him as a well-respected leader who tries in earnest to do right by his people and ensure that the assembly decides how tourism revenues are put to use for the benefit of the community at large. The returns from their strategic partnership with Treetops allowed Lolkisale village to invest in its own *maendeleo* (development) in a grassroots fashion, based on the needs of the community. Given Lolkisale's remote location, one of the key concerns voiced by community members in the context of development was the inaccessibility of health services. Road infrastructure in Lolkisale is poor, and this is made particularly challenging by black cotton soils, significant water runoff from Lolkisale mountain, and multiple rivers separating Lolkisale from the main A104 Arusha-Babati highway junction near Duka Bovu. While the road is relatively straightforward to drive in the dry season, it is treacherous in the wet season and becomes almost impassable without four-wheel drive.

Public busses are unfortunately unable to make the journey in some instances, and this can make life in Lolkisale dangerous when medical

Young men attempt to push a vehicle out of the mud on the road from Duka Bovu to Lolkisale. Photo by author in 2020.

emergencies arise. During my fieldwork in Lolkisale in early 2020, women were particularly concerned about the difficulties the road posed for reproductive health during pregnancy and childbirth. To address this issue, which disproportionately affected women in the community, LVC decided, with support from the assembly, to invest some of the initial profits from the partnership with Treetops in a four-wheel-drive community ambulance that could be used to shuttle people to the hospital during complicated childbirths or other emergencies. Fuel, maintenance, and vehicle repair costs were covered by the Treetops head fees, and the ambulance became a key fixture in Lolkisale village. Importantly, it was funded by conservation-related tourism, and not the government, despite providing an essential public service. Villagers in Lolkisale were aware of this fact and quickly came to realize the tangible benefits that conservation could have for their well-being. In a show of enthusiasm, the community printed "*maziwa ya tembo*" on the back of the vehicle, which translates into English as "the milk of elephants." In the same fashion that the Kisongo live off the milk of their cattle, the community of Lolkisale had come to realize that conservation could pay a regular dividend and allow them to flourish. It could take care of them in times of need and contribute directly to the economic prosperity of their community. Elsewhere in East Africa, Lesorogol (2022) has noted

An elephant walks in Randilen Wildlife Management Area with the Esimangore mountains in the background. Photo by author in 2019.

mobilization of this phrase by conservation NGOs in rhetorical terms that have not translated into significant trickle-down benefits at the local level. In Lolkisale, however, "elephant milk" was providing a crucial health service. Since the initial ambulance purchase, tourism revenue in Lolkisale has been used to build a health dispensary, a school, and several infrastructure projects in the village, which community members not only appreciate but depend on. A water tank was installed at Lemooti school, and desks and school supplies were purchased for students (King 2009). While cash itself did not trickle down to the household economies of families, it was used to create economic safety nets that insulated community members from the need to pay for community initiatives out of their own pockets. In thinking of tourism revenue as the milk of elephants, Lolkisale had come to realize that conservation could sustain their communal life. This was a deeply important sentiment because dispersing elephants caused significant problems for individual livelihoods and household economies.

As one of my interviewees in Lolkisale explained in early 2020, "elephants raid maize farms and affect our crop yields, but they also bring ambulances, and schools. Maybe one day, they will also help us build roads." In short, people in Lolkisale had come to realize that wildlife could be a sustainable resource pool for the village that would support the community and help it

A giraffe stands in Randilen Wildlife Management Area. Photo by author in 2019.

Wildebeest cross the plains near Sunset Hill in Randilen Wildlife Management Area. Photo by author in 2019.

prosper. Lolkisale had tasted the milk of elephants and had developed a thirst for wildlife-related tourism on village land.

The good governance practices of LVC in negotiating a fair deal with Treetops on behalf of the assembly, and in exercising transparency and accountability in determining how the funds would be applied at the village level for the betterment of the community, led to widespread support in Lolkisale for community-based conservation. These sentiments developed despite the shortcomings of the Boundary Hill joint venture and inspired the community of Lolkisale to conserve the LCA on their own accord. Sulle (2008) estimated that Lolkisale village earned $65,000–$78,000 per year between 2006 and 2008 as ecotourism on village land thrived. Lolkisale had become a bright spot for conservationists who were concerned about fragmentation outside Tarangire NP. As Rodgers et al. (2003, 11) optimistically noted, while the Tarangire ecosystem at large was "under increasing threat" of agricultural intensification, "Lolkisale [was] an important exception where conservation incentives created by wildlife-based tourism at the village level [were] reversing some of these land use changes." In short, the Treetops-Lolkisale payments model of community-based conservation was working.

CHAPTER THREE

Politics of Hunting and Photographic Tourism

While community-based conservation was showing promise in Lolkisale in the early 2000s, there were other political dynamics that complicated the situation. Following the Wildlife Conservation Act (WCA) of 1974, many of the wildlife dispersal areas adjacent to national parks were technically classified as Game Controlled Areas (GCAs). The Ministry of Natural Resources and Tourism identified Lolkisale as a key wildlife area adjacent to Tarangire NP and designated it as a GCA in 1974 to be managed by the Wildlife Division in collaboration with Monduli District Council (MDC). The GCA covered just over 1,000 km² and essentially spanned the entire eastern boundary of the park. Rather confusingly, the GCA overlapped Steyn's lease (chapter 2), which had been administered for commercial seed production and ranching. The abrupt cancellation of Steyn's lease in the early 1980s was likely due in part to this complicated dual classification of land. State willingness to offer leases to private investors like Steyn was conditional on the landholders maximizing productivity and generating profits for the central treasury. While commercial seed production had proved viable northeast of Tarangire NP, the abundance of wildlife in Lolkisale posed the dilemma of how best to generate capital in the area. As cautioned by Borner (1985), commercial agriculture along the eastern edge of the park was not only incompatible with wildlife-related economic activities outside the park but also threatened the state's ability to extract value from wildlife *inside* the park via the Tanzania National Parks Authority (TANAPA) by fragmenting critical seasonal wildlife habitat. The two potential resource uses were thus mutually exclusive, forcing the government to take a firm decision on which strategy it viewed as more productive. Ultimately, the government decided to favor the GCA and subsequently reallocated the land for trophy hunting. Bundu Safaris Ltd. held the Lolkisale block in the early 2000s and hosted tourist hunters in a small lodge called

Lolkisale Camp. Tourist hunters relished Lolkisale for its plentiful and approachable game, including big cats, buffalo, and elephants.

From the perspective of the central government, trophy hunting represents big money and is generally favored over other resource uses when the opportunity presents itself. Between 1989 to 2001, for instance, centrally collected revenues from trophy hunting increased sevenfold, highlighting state interest in supporting the sector (see Baldus and Cauldwell 2004; Nelson et al. 2007). Trophy hunting in Tanzania dates back to the colonial era, with some of the contemporary blocks remaining relatively unchanged since the 1950s. Following the WCA of 1974, trophy hunting became centrally managed by the Wildlife Division via game reserves (fully patrolled), GCAs (partially patrolled), and open areas (unpatrolled dispersal areas) (Wilfred 2019). Unlike game reserves, which prohibited all local economic activities, GCAs allowed livestock grazing, human settlement, and cultivation prior to 2009, but prohibited hunting without an official license. As discussed in chapter 1, GCAs enabled the central government to monetize wildlife resources that would otherwise remain outside the reaches of TANAPA, but their regulations were poorly conceived, and conflicts arose between layered land and wildlife laws (Wright 2016). On the one hand, Lolkisale GCA offered the Wildlife Division direct revenue from the hunting block. On the other, it was subject to encroachment and competing local interests. Rural communities, some of which expanded into the GCA and laid claim to land through cultivation following villagization, exercised customary land rights, posing a governance challenge for the state.

Since Steyn had been slow to develop his land for agriculture, other than some modest bean farms on the northeastern edge of Tarangire NP, much of the landscape was still open when the government canceled the lease (Borner 1985). Steyn had seemingly been weighing the opportunity costs of ranching versus seed production in the area. When the lease was withdrawn in the early 1980s, it catalyzed a scramble for land as smallholder cultivators, particularly Arusha, accelerated their migration into the area (chapter 2). Villages that were primarily demarcated during villagization based on preexisting Kisongo settlement patterns and the input from technical experts began to expand both in population and geographic sprawl. Eventually, sizeable sections of Lolkisale GCA overlapped the rapidly expanding human settlements, creating frictions between rural communities and state policy. While the southern parts of Lolkisale GCA remained open with no settlements or agriculture, smallholder cultivation rapidly expanded in the portion of Lolkisale GCA in Monduli District (Sachedina 2008; Sachedina and Trench 2009).

Between 1984 and 2000, Lolkisale GCA's landscape transformed from common pasture and small, scattered farms to extensive crop cover (Msoffe et al. 2011). This posed an issue to the Wildlife Division, which sought to monetize the hunting block. A series of land policies further complicated the situation. The Regulation of Land Tenure Act of 1992 attempted to eliminate conflicting customary rights to land altogether (Homewood et al. 2004; Homewood and Thompson 2010). Theoretically, this would have simplified things from the perspective of the state. In practice, however, people had already established residence in Lolkisale GCA and were difficult to move. When the Village Land Act and Land Act were reformed in 1998 (chapter 1), GCA inhabitants gained formal rights to remain provided that the settlement areas were designated as village land. The new acts provided village councils with the formal governance authority to manage and allocate land inside their jurisdictions, but the ambiguity around the GCA status was still unresolved. Lolkisale GCA came to directly overlap Lolkisale village in Monduli District (Arusha Region), and Loiborsoit, Emboreet, and Loibor Siret villages in Simanjiro District (Manyara Region). Conflicts between villages and the GCA quickly became entrenched. Emboreet, for instance, held that their village extended to the edge of Tarangire NP, but the central government considered that area to be part of Lolkisale GCA (Sachedina and Trench 2009). Sachedina (2006) even noted that Lolkisale GCA boundaries were contested between district (Monduli versus Simanjiro) and region (Arusha versus Manyara).

The Lolkisale Conservation Area (LCA) was originally proposed inside Lolkisale GCA, which was technically legal albeit confusing. According to the Australian investor's second loan proposal to GEF in 2003 (chapter 2), "the establishment of a Natural Resource Management Area within the Lolkisale Game Controlled Area and outside Tarangire NP is in accordance with the Wildlife Policy of Tanzania (MNRT 1998) and the Forestry Policy of Tanzania (MNRT 1998), which promote the conservation of wildlife, forest and associated habitats outside of core protected areas" (LBCSP 2003, 5). Included in the proposal was a letter of endorsement written by the Permanent Secretary of the Ministry of Natural Resources and Tourism (MNRT) with attention to the Director of Wildlife, clearly demonstrating that the central government was on board with the plan, at least in theory. The key issue with the overlapping villages and GCA was that Lolkisale village council had negotiated a village-based photographic tourism concession with Boundary Hill Lodge and Treetops but was technically dealing in land that was also being leased by the state to Bundu Safaris as a trophy hunting block. Both claims were legitimate to some extent, as different laws could be mobilized in support of either position. The question of whose rights should prevail in this situation has been an enduring one in Tanzania.

Discontinuities in the formal status of land in Lolkisale village and Lolkisale GCA created significant tenure complications, making it difficult for stakeholders to capitalize on wildlife in the area. The Tanzania Wildlife Management Authority (TAWA) and trophy hunting operators wanted the GCA hunting block to be prioritized, while Lolkisale village council and Treetops wanted the LCA and photographic concession to be honored. Though the Treetops agreement was more recent, it was layered on the preexisting GCA framework. In an attempt to streamline state policy, the MNRT passed a law in 2000 that prohibited photographic tourism, walking safaris, and game viewing in hunting blocks and GCAs (Nelson 2004; Rodgers et al. 2003). Exceptions would only be made with explicit permission from the Director of Wildlife since wildlife remained property of the state. As a consequence, the fruitful relationship between Lolkisale and Treetops essentially became illegal because the photographic concession overlapped the GCA. While MNRT's written letter of endorsement meant that the Treetops partnership with Lolkisale in the LCA could continue on an exceptional basis, it offered little in the way of political reassurance since it was technically an anomalous case outside formal law. It also further entrenched the contradictory nature of land and wildlife policies, adding to precarity on the ground.

Though the central government preferred to prioritize the GCA to generate capital for the state, this was not a straightforward proposition. The Lolkisale hunting block became harder to manage following villagization as opportunistic cultivators moved into the area, cleared large stretches of land for agriculture, and burned patches of forest for charcoal production. Institutional administration of the block was also challenging. Technically, Lolkisale GCA was managed by TAWA in collaboration with Monduli District. In accordance with the Wildlife Policy of 1998, hunting fees in Lolkisale GCA were collected by TAWA, and a quarter was then supposed to be returned to MDC. Of this quarter, 40 percent was to be allocated to the natural resources department of the district, and the remaining 60 percent was "expected" to flow down to local villages. According to the Monduli District Game Officer (Monduli DGO), this fee-sharing structure had been arbitrarily determined by TAWA without attention to trophy quotas or the number of animals killed (Sulle 2008). Furthermore, the funds never actually reached the villages in practice. TAWA retained the lion's share of these revenues, and there was little to no transparency in the transfer of funds to the district level.

Consequently, MDC began to develop resentment toward TAWA. MDC had seen its hunting revenues decrease in the early 2000s and suspected that the central government was taking more than its fair share (Sulle 2008). In response, TAWA representatives claimed that MDC was underreporting the funds it had received from TAWA and also pointed out that Lolkisale had low-quality game in its hunting block due to agricultural encroachment. The Monduli DGO, however, felt that trophy animals in Lolkisale were very high quality. He further alleged that TAWA had issued licenses without carrying out surveys of existing wildlife populations, insinuating that TAWA was milking wildlife for its own gain at the expense of the district government and the prospect of wildlife conservation in general. While I cannot substantiate these accusations, it is clear that the lack of institutional clarity around the revenue-sharing structure of Lolkisale GCA was causing confusion and distrust among government stakeholders operating at different scales. The Monduli DGO was particularly resentful because district game officers carried out the bulk of on-the-ground management duties. While TAWA game officers did patrol the GCA, they did so haphazardly, given the numerous other GCAs and game reserves across Tanzania that also needed to be monitored. From the Monduli DGO's perspective, TAWA used Lolkisale GCA to grab wildlife revenue without offering adequate support to the district level to actually administer it. MDC had far less funding than TAWA to pay staff and fuel vehicles, constraining district management practices and engendering feelings of bitterness toward TAWA.

Resident hunting further complicated territorial conflicts between trophy hunters, photographic operators, and Lolkisale village in the early 2000s. Tanzania's resident hunting program was established in 1974 and banned between 2015 and 2018 following widespread allegations of corruption and abuse. The government introduced it to allow local consumptive wildlife uses in areas outside reserves and parks, in either open areas or GCAs (Wilfred 2019). District councils oversaw the administration of licenses under the directives of TAWA, but the system faced numerous issues including noncompliance with regulations, illegal photocopying and sharing of permits, violation of quotas, underreporting kills, reusing licenses, and targeting prohibited animals (Baldus and Cauldwell 2004; Wilfred 2019). Furthermore, local communities were usually unable to access resident licences due to high fees and centralized control of processing (Leader-Williams 2000). Consequently, wealthy elites from afar often financed operations, undermining the notion that the program supported "resident" hunters. Open areas were also difficult to enforce in practice due to limited management capacity, creating favorable conditions for poachers to capitalize on the inconsistent surveillance (Caro and Davenport 2016).

A key issue in Lolkisale stemmed from the enforcement of hunting regulations on the ground and conflicts between resident hunters and trophy hunters. Technically, resident hunters were prohibited from accessing the GCA block that was allocated to trophy hunters, though they could access open areas nearby. However, boundaries were poorly marked and difficult to enforce in practice across a large area. Trophy hunters felt that resident hunters encroached on their designated block and disrespected local hunting regulations. Drawing from cases elsewhere in Tanzania, Wilfred (2019) suggests that these are common allegations raised against resident hunters, who are sometimes beaten and harassed by trophy hunters. Resident hunters commonly complain in response that trophy hunters encroach on open areas in order to bait carnivores. A lack of consistent monitoring and enforcement on the ground by either district government or TAWA meant that conflicts between resident hunters and trophy hunters were frequent in Lolkisale in the early 2000s.

Trophy hunters and photographic tour operators alike agreed that resident hunting east of Tarangire NP was poorly regulated and associated with widespread poaching. In the 1980s, there was a poacher's camp in Lolkisale village, near Boundary Hill, though poaching dropped off considerably after 1989 following the international ban on ivory trade (Pittiglio et al. 2013). Despite the declines, conflicts in Lolkisale over access to, and use of, wildlife resources had become entrenched. Multistakeholder tensions had arisen between TAWA, MDC, resident hunters, trophy hunters, photographic tour operators, and Lolkisale village. These conflicts were largely attributable to layered national policies, which created unnecessary ambiguity and overlapping jurisdictions of formal rights. Enforcement of laws on the ground was complicated by decentralization of the state into different arms of MNRT, district councils, and village-level governments, all of which were constrained by limited management capacity in practice.

Contestations over Lolkisale GCA

Conflicts over Lolkisale GCA/LCA continued throughout the early 2000s and, to resolve them, a series of multistakeholder meetings were held in 2008. During one meeting in May 2008, representatives of trophy hunting companies, photographic operators, a conservation NGO, and the Wildlife Division gathered to address escalating tensions over the use of wildlife resources in Lolkisale GCA/LCA. I provide some thick description of the meeting in the following paragraphs based on publicly available minutes published online by the Tanzania Natural Resources Forum. A key topic of discussion at the meeting was a proposal to implement no-hunting buffer zones around Naitolia Camp, Treetops Lodge, and mobile camp sites used during walking safaris (TNRF 2008). For safety reasons, trophy hunters and photographic tour operators could not operate in the same areas at the same time. In Lolkisale, hunting on foot was generally not feasible because the concession was so large. Generally, hunters needed to "drive, spot, and stalk," often taking shots from about 75–250 yards away. This made it particularly dangerous for tourists to stroll around the hunting blocks as part of walking safaris. The question of whose access and use rights should take precedence in such instances had led to growing hostility between hunting and photographic tour operators.

The Australian investor's brother advocated at the meeting for an expanded buffer zone around Tarangire NP of 4 km. His own interests notwithstanding, he framed his proposal as a means of ensuring that wildlife populations were protected and that local communities could reap some benefits from wildlife through private contracts with tour operators. MDC had indeed previously agreed to increase Tarangire NP's buffer zone to 5 km in real terms, but they could not codify it into law because the district did not have the authority to do so without approval from the Wildlife Division. While the 2 km buffer zone around the park was already in place, and most ecotourism camps were located beyond 3–4 km from the park, Treetops was located between 2 km and 4 km from the eastern boundary, in the same area as the trophy hunting concession. Most meeting participants expressed skepticism that the central government would agree to such a bold proposal, so the Australian investor lobbied instead for the standard 2 km buffer zone around the park to be respected, with "no-take" circles added around Naitolia Camp and Treetops Lodge. Boundary Hill was excluded from the proposal, as it had recently burned down. The problem, however, was that the additional buffer zones would cut into hunting Block B within Lolkisale GCA, so Bundu Safaris stood to lose from the change. Notably, Bundu Safaris did not have a representative at the meeting.

An associate from Treetops appreciated the proposal but was unconvinced about how the added loops would be enforced. Practically speaking, nothing could be put into law prior to the expiration of the current lease contracts. While the Australian investor wanted the Wildlife Division to swiftly implement the expanded buffer zone through legislative reform, other attendees recognized that this could take years. This consideration raised the question of how the conflicts would be resolved in the interim. To quell the concerns from Treetops's representative, one tour operator explained that the agreements reached during the meeting "will be the law for two years," implying that since the meeting attendees were the resource users in practice, if they considered the rules to be binding, then they would be (TNRF 2008, 3). Interestingly, then, the meeting signified the creation of an informal institutional arrangement for governing access to wildlife resources in the LCA/GCA dispersal areas outside Tarangire NP that unfolded in the cracks of formal state legislation, much in the same fashion that common property mechanisms operate at the community level through custom.

The Treetops representative expressed further concern that because Bundu Safaris was not represented at the meeting, the proposal would likely not be respected. To this, another operator pointed out in response that the agreement would serve as a "consensus-based solution" determined by the majority of companies operating in the area (TNRF 2008, 4). Other meeting participants, however, agreed that political backing was a necessary prerequisite for enforcement. One individual lodge owner mentioned that the Wildlife Division was well aware of the emergent conflicts and was eager to find an "amicable solution" to resolve them (TNRF 2008, 4). As he noted, the Wildlife Division was prepared to mandate different zoning arrangements in 2009–2010 to formalize a compromise between hunters and photographic tour operators but conceded that the meeting participants needed to implement a short-term solution until then. Ultimately, the group voted on the size of the buffer zone and agreed to extend it an extra 2 km all the way around the park (4 km in total). The expanded buffer was never formalized as law but was still put into practice on the ground.

To alleviate potential conflicts between Bundu Safaris and Treetops, one third-party operator suggested allowing Bundu to continue hunting between 2 km and 4 km of the park because its hunting block was fairly small. The Treetops representative pushed back, knowing that his lodge, walking safaris, and fly camping operations were at stake, though his counter proposal was framed in terms of ecological reasons. Others agreed that the expanded buffer zone would likely benefit the environment and could also be combined with village-based photographic concessions to further benefit communities. The question of how the Wildlife Division would get its cut of these arrangements, however, was raised as a potential issue, and the group suggested a "wildlife activity fee" to be paid to TAWA on top of the concession fees paid to the contracted village (TNRF 2008, 6). One operator then firmly clarified that this should not constitute a "double payment" for the concession, but an additional flat rate tax so that the state would also benefit from the arrangement (TNRF 2008, 6). The group also decided through verbal agreements that photographic operators would inform hunters of their mobile camping schedules in advance to ensure the safety of tourists. Hunters would then have to refrain from hunting within 1 km of the mobile camps, even if the blocks overlapped the mobile photographic camps.

The inter-stakeholder governance meetings between tour companies operating east of Tarangire NP reveal that there was an institutional scramble underway to harness the capital generated by wildlife resources on village land. This involved a combination of formal and informal governance and management arrangements coexisting at different scales. The lack of formal harmonization in state policies had resulted in competing claims that were supported by law, and this made it challenging for all involved stakeholders to ensure consistent access to the benefit streams of wildlife resources. From the perspective of the central government, wildlife was state property and was a key source of revenue through TANAPA inside Tarangire NP, and TAWA in Lolkisale GCA. From the perspective of Lolkisale village, wildlife was passing through village land, over which village councils had formal authority. Thus, the lack of clarity over governance of wildlife resources led not only to conflict but to strategic alliances: the central government sided with trophy hunters, and Lolkisale village positioned itself with the photographic tour operators.

Legislative Reform

Indeed, as one tour operator had predicted during the inter-stakeholder meeting of 2008, legislative reform was just around the corner. The revised WCA No. 5 of 2009 addressed many of the ongoing grievances that had developed about GCAs and villages (chapter 1). The new legislation ultimately clarified that village land could not overlap GCAs, and in such instances where it did, settled areas would be classified as villages pursuant to the Village Land Act (PINGO 2013). The new WCA also prohibited livestock grazing, cultivation, and settlement inside GCAs, eliminating some of the prior confusion: villages were for communities, and GCAs were for centrally managed trophy hunting concessions (MNRT 2009; Wright 2016). This greatly clarified the situation, as large parts of Lolkisale GCA became superseded by village land. The Revised WCA of 2009 built upon the legislation of 2000 by further restricting private contracts between village councils and tour operators that had been negotiated without state approval. The main reason for this was that there was no legislative framework in place for centrally taxing the wildlife-related revenue streams that accrued from them. The new act thus demonstrated a clear attempt by the state to reconsolidate central control over wildlife revenues. Neoliberal partnerships between villages and private investors that had emerged in the context of community-based ecotourism in the 1990s (cf. Gardner 2016) were thus short lived, as the state once again intervened in the market by implementing regulatory policies from above. As described in chapter 1, this was the invisible hand behind the concept of WMAs in Tanzania, which represented a formal institution for extracting taxes from wildlife-related tourism occurring on village land pursuant to the Wildlife Policy of 1998. The revised WCA of 2009 clarified that all tourism contracts with safari operators on village land *had* to be implemented within the framework of a WMA. Consequently, Lolkisale village was forced to decide whether to establish a WMA or forgo their productive agreement with Treetops and the prospect of a community-based livestock and wildlife area altogether.

In forwarding WMAs as the official model of community-based conservation on village land, the state was not acting in isolation from other influences. The African Wildlife Foundation (AWF) had begun to pressure the central government in the early 2000s to push the WMA model. AWF touted WMAs as part of its "heartland" initiative, which sought to conserve key ecosystems outside national parks across Tanzania's Maasailand (Bluwstein 2018a; Gardner 2016; Igoe 2010, 2017; Wright 2019, 4, 102). In the northern parts of Maasailand, AWF referred to its program as the "Kilimanjaro Heartland," which extended from Enduimet across to Lake Natron and covered most of the Longido District (Wright 2019, 108). In the Tarangire ecosystem, the area was referred to as the "Maasai Steppe Heartland."

AWF strategically fashioned its regional program in relation to the emergent global interest in landscape-level conservation and the emotional appeal of American heartlands (Sachedina 2008). To attract donor funds from USAID and the World Wildlife Fund (WWF), it carefully crafted a marketable image of conservation in the Tarangire ecosystem (Igoe 2017). AWF's websites and brochures highlighted pastoral Maasai living harmoniously together with wildlife while downplaying the presence of Arusha cultivators and other social complexities that disrupted its aesthetic of wildlife conservation in Maasailand (Igoe 2010). In so doing, AWF was able to establish itself as the dominant NGO in the regional conservation arena in the 2000s, vacuuming up much of the global donor funding in the process (Bluwstein 2018a, 2018b). At the suggestion of AWF, the central government commenced a WMA "pilot period" from 2003 to 2006, where WMAs were implemented on an experimental basis throughout Tanzania (Kimario et al. 2020, 127; Wright 2019). With AWF's technical and financial assistance, Enduimet WMA was conceived in 2003 and gazetted in 2007; Makame WMA was introduced in 2003 and formalized in 2009; and Burunge WMA was initiated in 2003 and formally established in 2006 (WWF 2014). AWF's efforts were driven by an enduring wildlife-preservation ethos. My own interviews with George Sanford (pseudonym) in 2020, who served as AWF's Maasai Steppe Landscape Director during its heartland initiative, revealed a genuine interest in protecting wildlife from poaching and safeguarding the habitats upon which they depended. Lolkisale had been identified by AWF as an important corridor out of Tarangire NP and a poaching hotspot, and was thus high on its agenda for a WMA (Bluwstein 2018b, 160). Notably, AWF was generally opposed to trophy hunting and viewed Lolkisale's interests in photographic tourism more favorably than the GCA hunting block.

From Sanford's perspective, AWF was doing well-intentioned work in Maasailand to protect vulnerable and endangered species of wildlife. The organization inspired a large collective of African practitioners and researchers who cared about the well-being of wildlife, largely through the College of African Wildlife Management Mweka, which was founded with a small grant from AWF in 1963. As Sanford described to me, central to the heartland scheme was a focus on protecting wildlife that moved onto village land and would otherwise be unmonitored by national park rangers. AWF was certainly not blind to the concerns of communities, but its ultimate priority was conserving wildlife. According to Sanford, AWF latched onto WMAs as the best tool available for wildlife conservation on village land because the heartlands program still had to operate within the framework of national policies for governing land and wildlife resources. WMAs were one of the only viable options available for an organization that had secured considerable donor funding and was eager to contribute to wildlife conservation outside national parks; WMAs could hypothetically appeal to government, community, and investor interests while achieving AWF's ultimate goal of conserving wildlife on village land. Thus, AWF likely meant well by trying to rapidly establish WMAs in wildlife dispersal areas throughout Maasailand. At the same time, it is difficult to overlook AWF's impacts on the Tarangire ecosystem from a social perspective (Bluwstein 2018a; Igoe 2017). In the mid-1990s, for example, AWF resettled sixty people living to the northwest of Tarangire NP to make way for a wildlife corridor that would later become Burunge WMA (Bluwstein 2018b; Igoe and Croucher 2007). Based on his time spent working in a prominent role with the organization, Sachedina (2008) suggested that AWF's conservation efforts tended to benefit select elites while stoking resentment at the community level.

While the Wildlife Policy of 1998 stipulated that WMAs could be established on village land in key wildlife habitat areas adjacent to national parks, it offered little insight into how they would actually be operated (Nshala et al. 1998). Rural communities perceived WMAs with distrust and often resisted them during the initial pilot phase (Bluwstein et al. 2016). The procedural processes for establishing WMAs were convoluted and unclear, and communities simply did not understand what they would entail in practice (Nelson et al. 2006). Establishment of WMAs also tended to drag out over several years (Rodgers et al. 2003). In Lolkisale, the Kisongo were mindful of their history of dispossession from Tarangire NP and the politics of Lolkisale GCA. They were wary of a prospective WMA, which they initially viewed as a government scheme. There was concern that Lolkisale village might lose its rights to allocate land and authorize local tourism investments. The initial WMA legislation also failed to stipulate exactly how much revenue local communities would receive from tourism as compared to TAWA. Lolkisale village council was worried that high taxation would eat significantly into its revenue stream, potentially altering the opportunity costs of keeping land open for photographic concessions.

Realizing that its arrangement with Treetops was legally precarious and in need of formal reclassification, Lolkisale village council weighed the potential risks and benefits of establishing a WMA. On one hand, the WMA was an unknown entity that was devised by the central government and thus posed significant risks to the village, both in terms of potential revenue losses and land dispossession. On the other hand, the LCA was constrained by a lack of political backing, making it difficult to enforce their livestock area and photographic tourism concession. The Lolkisale council was also worried about the government deciding to cancel their contract with Treetops on the grounds that it did not exist in accordance with state law. Thus, the proposition of a WMA in Lolkisale village had potential trade-offs and, at first, the community was unsure of how to proceed. In 2010, Lolkisale village entered into discussions with AWF and the Monduli DGO about the possibility of registering a WMA in Lolkisale village, at which point district government officials explained that the WMA would have to include other neighboring villages as well. This added even more nuance to the decision, as bringing other villages into the fold could give rise to new political complexities and potentially jeopardize Lolkisale's authority over its tourism arrangements.

The Kisongo of Lolkisale were particularly intrigued by the potential for formalization to increase their capacity for territorial defense. While they had appreciated the LCA, its integrated management plan had been difficult to enforce in the face of competing interests in the area from hunters and commercial farmers. Even after the legal reform, trophy hunters still patrolled the areas deemed to be part of the hunting block, as did resident hunters and poachers. The community of Lolkisale resented the fact that hunters were entering their territory and undermining the resource base of their lucrative arrangement with Treetops. Though the revised WCA had made clear the distinctions between village land and the GCA in terms of formal tenure, in practice, there was driving concern as to who would enforce these laws on the ground. TAWA and the Monduli DGO had limited capacity for monitoring the area, and the local Kisongo were not equipped to confront hunters. Community members complained as well that trophy hunters and illegal poachers beat herders if they interfered with hunting activities, even if hunting was taking place on village land (Nshala et al. 1998). By rezoning the area as a WMA, the livestock zone would officially be endorsed by TAWA, and a team of village game scouts could be brought in to patrol the area and enforce pastoral territory. As outlined to me during an interview with a Kisongo man (about forty-two years old) in Lengoolwa:

> The original idea to establish a WMA was from the leaders of Lolkisale. Before the WMA, there was only Treetops over there, but it was a Game Controlled Area. There were a lot of hunters and poachers. Of course, wildlife can be destructive and stubborn, but we did not like seeing people kill them, especially since we were getting money from tourists coming to stay at Treetops. If a hunter came, he just went straight to that GCA, hunted and left. We did not know where the hunters were coming from. So our village leaders asked: "Why are these people just entering our village, hunting and leaving without even reporting to us? We need community control of this area, so why don't we form a WMA."

Further to hunting, the Kisongo were also becoming increasingly concerned with the commercial farmers who had moved into Lolkisale after the cancellation of Steyn's lease. The LCA's integrated management plan offered little guidance on how these farmers could be pushed out of the area. This was a particular concern in Lolkisale's Lemooti sub-village, which was a key area for grazing and was becoming enclosed by several large commercial farms. Importantly, the farms were located inside the LLWZ area, which the Kisongo resented, especially since the farmers were mainly Arusha, Meru, and Chagga elites who lived in Arusha town and Moshi. In the mid-2000s, the issue of absentee landholders had become a major problem in the Lolkisale area and was putting further land pressure on the pastoral Kisongo. While fraught with risks, the Kisongo reasoned that formalizing their community-based conservation area through a WMA might be a means of destabilizing the farmers' claims to land, especially since this approach would have the support of the central government.

The Ripple Effect of WMAs in Maasailand

The growing concerns among the Kisongo about hunters depleting their profitable resource pool (wildlife), and Arusha farmers grabbing pastoral land, were strong incentives for Lolkisale village council to entertain the idea of upscaling the LCA to a WMA, despite their worries about the risks involved. While the Kisongo recognized the potential benefits that the WMA could have, they were wary of potentially committing to something that could have significant consequences for the community in the long-term. Some elders were concerned about the lack of legal precedent for villages to leave WMAs once they had been formalized (chapter 1). To determine whether they truly wanted a WMA in their village, the *korianga* (junior elder) age-set of Lengoolwa sub-village organized a visit to the Kisongo of Longido, where Enduimet WMA had been established in 2007. I interviewed members of this original party in early 2020, and they explained with great pride how they had held a series of meetings in Lengoolwa in 2010 to discuss the idea of the WMA before traveling. After selecting a key group of scouts, they raised money on their own through village contributions to fund the trip to Longido in 2011. Delegates from the neighboring villages of Makuyuni, Naitolia, Mswakini Juu, and Mswakini Chini were invited to attend based on the district government's suggestion to band villages together. Five representatives were selected from each of those villages to visit Enduimet WMA and observe it in practice. Also accompanying them on this trip were ward councilors, village chairs, and district representatives. The group was led by several *ilaigwenak* (Maasai traditional leaders) who sought out respected elders in Longido to discuss their experiences with Enduimet WMA. They were tasked with a mission to deliver the messages of these elders back to the community of Lolkisale. As documented by Wright (2019), the Kisongo of Longido had also wrestled with the question of whether a WMA would be in their community's interests. At first, they opposed the prospect of a WMA, worried that it would displace their land rights. Over time, however, they came to view it as an opportunity to strengthen sovereignty over their traditional territory in the face of trophy hunting and foreign investors they did not care for (Wright 2017).

To fully understand what was garnered from the meeting with traditional leaders in Longido, it is crucial to understand the ethnographic context of Enduimet WMA, which was one of the flagship WMAs during AWF's initial pilot period. Like Lolkisale, Enduimet comprised a Kisongo majority spanning nine villages in Longido District. Wright (2017) described three main events that galvanized community support for the WMA after it was established in 2007. First, in an eerily similar fashion to the political dynamics in Lolkisale village (and Loliondo), Sinya, the largest village by area in the WMA, had been struggling to secure photographic tourism on village land that overlapped a trophy hunting block in Longido GCA. Sinya village was located just across the border from Amboseli National Park in Kenya and thus constituted an important dispersal area for elephants moving outside the park (much like Lolkisale and Tarangire NP). As was the case in Lolkisale, the area was dual classified as the Longido GCA, and the block was leased to a trophy hunting company named Northern Hunting (Trench et al. 2009; Wright 2016, 2017, 2019). In the late 1990s, Sinya village began to welcome photographic tourism on village land but ran into a similar conflict as Lolkisale village because their concession overlapped the one that had been leased by TAWA to Northern Hunting (Wright 2017). The community took Northern Hunting to court, and the central government unsurprisingly sided with the trophy hunters. While Sinya village had initially been opposed to the notion of a WMA on their land knowing that it represented a tool for centrally managing wildlife revenues, they abruptly changed course when the WCA was revised in 2009 (Wright 2017). Sinya village council realized that by joining Enduimet WMA, they gained the power to evict Northern Hunting and re-assert sovereignty over pastoral territory. Through the WMA, Sinya was able to rezone the area for photographic tourism and push out the hunting concession. With Sinya's interests represented, the Authorized Association (AA) of Enduimet WMA revised its management plan for the period of 2011–2016, and in 2012, the Director of Wildlife approved the new zoning plan—the trophy hunting block had effectively been zoned out (Wright 2017).

Not unlike the controversies that had arisen in Lolkisale with Boundary Hill Lodge, Sinya village had also run into issues with a photographic tour operator. In 2007, Shu'mata Camp established itself in Sinya through a nefarious arrangement with former village leaders who received personal handouts in exchange for the rights to an exclusive tourism concession (Wright 2017).

The investor received a written letter of endorsement from the villager chair, which sufficed as proof of formal allocation of land pursuant to the Village Land Act. Herders in Sinya felt that the deal was made in bad faith, and the majority wanted the investor expelled. Despite attempts from the subsequent village council to create a new contract, the investor was unreceptive. Shu'mata abused its position in the community, offering no compensation for wildlife impacts on livelihoods, no employment opportunities, and no access to the tourism revenue streams (Wright 2017). Through the AA governance meetings of Enduimet WMA, the community decided that Shu'mata had to leave, and the community went as far as blocking roads, hijacking vehicles, and storming the lodge (Wright 2017). The investor sought support from politicians and other elites to crack down on the community but was eventually hit with an eviction notice in 2014 from the district government. Shu'mata took the village to court later that year, and after a two-year saga, the High Court ultimately sided with the community. The investor appealed the court decision, and the case was still ongoing as of 2020 but has since been dropped for reasons not entirely clear (Wright, personal communication, 2022).

This backdrop of conflict between the Kisongo of Sinya and foreign investors is crucial context for understanding what lessons were learned from the scouting trip to Enduimet WMA in 2011. Since the Lolkisale party had traveled to Longido two years after Sinya had changed course, from opposition of the WMA to support of it, the group took away from the visit an appreciation of the benefits that could be gained from establishing a WMA in their village. Sinya was in the midst of ousting the trophy hunters through their revised management plan, and thus the message that they delivered to the Kisongo of Lolkisale was very clear: WMAs were political tools that could be used to harness the power of the central government to push out unwanted outside interests in pastoral land and the wildlife it supported. The community of Lolkisale was particularly pleased with the idea of zoning out trophy hunting altogether. Quite significantly, then, the lived experiences of the Kisongo in Enduimet WMA had a ripple effect on the perspectives of the Kisongo in Lolkisale toward the prospect of a WMA.

The travel party members returned to their respective villages to report their findings, at which point local governance at the village level came to play a crucial role in determining the trajectory of community sentiment toward the prospective WMA. In Lolkisale, open village meetings were held, and the findings were discussed in a transparent way with the assembly. The greater Lolkisale community (including all five sub-villages) contemplated the potential benefits of joining the WMA in the context of their fruitful arrangement with Treetops and their tensions with Bundu Safaris and the central government over the hunting block in Lolkisale GCA. The experience of Sinya village in successfully pushing out hunters was highly reassuring, as were its plans to expel the problematic photographic tour operator. The Lolkisale community viewed these as important reassurances that the WMA could be used to serve their interests. But some key council members raised concerns during the assembly meetings about the impacts of institutional change on Lolkisale's revenue from Treetops. How much "milk" would be left for the community of Lolkisale? Would it become spattered across other villages or be slurped up by the central government? This proved to be a resonating concern that was difficult to address in the planning stages because the exact revenue structure of the prospective WMA was not yet known to the community. The village ultimately reasoned that the milk of elephants was something that would support their community indefinitely as long as wildlife populations remained healthy. Even if the revenue from their private arrangement with Treetops decreased, it would perhaps become a more reliable and secure long-term income stream with the added protection afforded by the WMA. Furthermore, pastoral livelihoods would be secure as long as the rangelands were not enclosed by commercial farms and Arusha encroachment. Formalization offered a political instrument for defending community land from the hunters who threatened wildlife and the farmers who fragmented habitat and reduced available pastures. Thus, the potential losses in revenue were seen as secondary to the long-term benefits of sustainably managing the two resources that were central to the community's well-being—land and wildlife.

CHAPTER FOUR
Creating a Wildlife Management Area

As the political leadership of Lolkisale contemplated the process of upscaling the Lolkisale Conservation Area (LCA) to a Wildlife Management Area (WMA), another social dimension featured into WMA preplanning: intervillage politics. The prospect of forming a coalition of neighboring villages through the proposed WMA had potential benefits and costs for Lolkisale. On one hand, it allowed the community to safeguard a larger area, but on the other, it meant that they would also have to share governance authority. This had the potential to put Lolkisale's private contract with Treetops at risk and weaken the pastoral values underpinning the assembly's interest in securing a community-based livestock area. The logical villages to include in the WMA, other than Lolkisale, were Makuyuni, Naitolia, Mswakini Juu, and Mswakini Chini. These villages shared borders with the LCA and had contributed land to the Tarangire Conservation Area through the elephant dispersal area in Makuyuni and Naitolia Camp. Together, they geographically surrounded the conservation area in question, and each village council had already expressed interest in community-based ecotourism on village land. Lolkisale's village council, however, was wary of these villages, largely because they comprised strong Arusha majorities. Reflecting on these apprehensions when asked about the process of WMA planning, one Kisongo woman from Lemooti (about sixty years old) remarked during an interview, "The big concern for us in the early stages was that the Arusha wanted to take our whole grazing area for farming!" As discussed in chapters 1 and 2, the Arusha had been rapidly expanding into the area, and their political and economic concerns differed from those of the Kisongo. In Lolkisale, the Kisongo still held cultural hegemony over the area, with Arusha residing in small, growing hamlets in Lengoolwa, Nafco, and Lolkisale-proper. Kisongo individuals still held sub-village chair positions in Lemooti, Lengoolwa, and Oldonyo. Nafco, however, was developing into an emergent Arusha-dominated breakaway sub-village, with Kisongo inhabiting only the peripheral areas away from the clustered farms. By way of contrast, villages on the Makuyuni side of the proposed WMA had become Arusha strongholds. Makuyuni had an Arusha majority excluding Saburi sub-village, and Naitolia, Mswakini Chini, and Mswakini Juu were inhabited almost entirely by Arusha. Those villages thus had different livelihood priorities than Lolkisale: their primary concern was smallholder farming, with livestock a secondary consideration. Since they were relatively recent migrants to the area, from the 1950s onward, the Arusha were also worried that their "indigenous" rights to the area would be called into question by the proposed WMA. An Arusha elder (about seventy-seven years old) from Lengoolwa described these ethnic frictions to me during an interview:

> Of course, there was a time when people quarreled about whether to establish the WMA. Some of us wanted even to shed blood over of this conflict. Because the Maasai and Arusha are different. The Maasai farm but they are not really farmers. We Arusha are farmers. We Arusha were afraid this place would become like Ngorongoro and prohibit crop cultivation. So even when the Maasai decided they wanted the WMA, we Arusha told them please, do not agree to things so easily. We have to know first that we will still be able to keep our farms.

While Lolkisale's village chair was himself a Nyaturu, he recognized that the interests of the Kisongo in securing a livestock area and promoting photographic tourism served the community as a whole irrespective of ethnicity. The village councils of the other four prospective villages, however, were held by Arusha, and there was significant concern within Lolkisale that including those other villages would sway the balance of power over the area in favor of a cultivation-based regional economy and increase encroachment of Arusha on pastoral land.

While the Lolkisale assembly did not yet understand the nuances of what a WMA would entail, they were quick to key in on the potential for other villages to eat into their income stream from Treetops. More importantly, they were concerned that the material base of their entire economy was at stake if the institutions for governance were to fall into the hands of Arusha-dominated councils that did not share their land use concerns. As one of my interlocutors in Lolkisale explained to me, this was a discreet factor underlying Lolkisale's process of subdivision. After discussions with the African Wildlife Foundation (AWF) and the Monduli District Government about what the WMA would involve, there was an awareness within Lolkisale that each member village would have an equal seat at the table. As things stood in 2010, this meant that Lolkisale, Makuyuni, Naitolia, Mswakini Chini, and Mswakini Juu would be on equal footing in governing the area, and thus, Lolkisale would be greatly outnumbered. Of particular concern, Arusha would hold 75 percent of the governance authority over the WMA and would receive that same amount of whatever tourism income was returned to member villages. Deeply troubled by this prospect, the Lolkisale community devised a plan to subdivide Lolkisale village into several smaller villages to increase its representation in the proposed WMA. Lolkisale was aware that Naitolia and Mswakini Juu had both subdivided from Mswakini and considered subdivision a useful strategy for increasing their political representation and economic stake. While it is common for villages in rural Tanzania to subdivide as their populations grow, the swiftness with which Lolkisale subdivided not only into two villages, but into *five*, was notable. In 2012, Lolkisale split into Lemooti and Nafco, and later, in 2015, Lengoolwa and Oldonyo followed suit. These processes took some time to complete, as they required mapping out village boundaries and registering them with the district, but all things considered, they were formalized quite quickly. The plan to subdivide Lolkisale into separate villages gave the assembly more confidence in the idea of establishing a shared WMA with Makuyuni, Naitolia, Mswakini Juu, and Mswakini Chini because Lolkisale would increase its political representation from 25 percent to 56 percent, allowing them to retain a governance majority. It would also bring three Kisongo-led villages to the governance table in Lemooti, Lengoolwa, and Oldonyo to balance the Arusha-led ones on the other side.

While the exact revenue structure of the WMA would not become clear until the revised WMA guidelines of 2012 were published by the Tanzania Wildlife Management Authority (TAWA), the math for calculating how this would affect income-sharing between villages was simple enough for the Lolkisale community to grasp. Rather than Lolkisale village receiving a combined quarter of the total income that would then be divided across its sub-villages, each sub-village would directly receive a ninth of the total income. Hypothetically, if $40,000 was returned to the community from the WMA, this would have initially been split four ways, meaning $10,000 for Lolkisale to divide among its sub-villages ($2,000 per sub-village). Comparatively, if Lolkisale subdivided into five villages, the $40,000 would be divided across all nine villages, meaning $4,444 each and $22,222 total for the Lolkisale community at large. Though not the only factors influencing the decision to subdivide, the political and economic motivations were noteworthy as Lolkisale strategized about how to navigate the WMA framework to protect its interests.

In the end, the community of Lolkisale decided to move forward with the WMA. Transparent governance institutions spearheaded by the village's widely respected chair ensured that the community was on board with the decision. A Kisongo woman (about seventy-five years old) from Lengoolwa described this process to me from her perspective when asked about Lolkisale's decision to form to a WMA: "I would not say that we had a conflict here about whether to establish a WMA or not, but we carefully considered whether it would be in our community's interest. We have good leaders here and had lots of discussion and people agreed together to form the WMA." Villages on the other side of the WMA, however, arrived at different conclusions. Following a series of community meetings, Makuyuni decided that it did not want to participate in the WMA, despite the continued interest of the village council in being included. Consequently, Makuyuni was dropped from the proposal and was ultimately excluded from the WMA.

There were several reasons why the community of Makuyuni decided not to be part of the WMA, some of which were sociocultural and others that were political and economic in nature. The primary one was that Makuyuni comprised an Arusha majority. Excluding the town, the sub-villages south of the A104 highway were inhabited mostly by Arusha who had in-migrated to the area during the socialist period. The Arusha of Makuyuni were concerned that the WMA might be used as a way for the pastoral Kisongo to prioritize livestock grazing at the expense of farming or, worse yet, to push the Arusha out of the area completely. As one Arusha man (about sixty-one years old) from Makuyuni said during an interview in 2019, "Makuyuni was one of the first villages in the WMA proposal, but we refused it because we worried that the land was going to be taken by the WMA and barricaded. We were concerned that our farms would be confiscated, and that the area would be kept only for grazing." Makuyuni residents were aware that the Kisongo resented Arusha in-migration and saw the WMA as a threat to their settlements and farms. Arusha elders were particularly apprehensive in this regard. Once a consensus had been reached around the risks that the WMA posed, a well-respected group of Arusha elders advised the assembly that if anyone decided to support the WMA against their community's wishes, there would be consequences. Some of the elders placed a curse, which translates roughly into English as a "broken pot curse" (*kuvunja chungu*—"to break a pot"). The broken pot curse is a traditional form of Arusha sorcery used to convince people to follow someone's orders. As part of the curse, pots are imbued with social life. If someone violates the terms of the curse, the elder breaks a pot, symbolizing the shattering of the cursed individual's life. When the pot is broken in this fashion, the individual and his or her family are destined to die abruptly. The curse extends to kinship ties and all descent lines from the individual that come after. One notable elder said to the community during an assembly meeting, "If you support the WMA, the pot will follow you and you will break. Your family will die as this pot dies." The broken pot curse is considered one of the gravest forms of sorcery at hand for the Arusha, so when people were faced with the options of being cursed to death or withholding support for the WMA, most people heeded the warnings of the elders.

Another reason that Makuyuni decided not to join the WMA was that an influential member of the Ujamaa Community Resource Team (UCRT) lived in the Kisongo-dominated Saburi sub-village. As discussed in chapter 1, UCRT was critical of WMAs because it viewed them as an extension of the central government that could be used to grab pastures away from pastoralists on village land. UCRT had worked diligently to ensure that after Steyn's lease was canceled in Makuyuni, the land was turned over to the village to be managed as communal pasture. *Ndoroboni*, the communal grazing area adjacent to the Makuyuni Elephant Dispersal Area (MEDA), was a crucial lifeline for the livestock keepers of Makuyuni, who were being squeezed on all sides by changing land tenure arrangements. The proposed WMA was going to overlap *ndoroboni*, so the Kisongo of Makuyuni, with input from UCRT, worried that this would mean losing control over their vital grazing area. Thus, while the Kisongo of Saburi and Lolkisale both prioritized communal grazing areas, they had different views on whether the WMA would help to secure or undermine them. In the end, the Kisongo of Makuyuni also felt that the WMA posed too great a risk to their village's grazing area and opposed the proposal to join the WMA.

The final reason that Makuyuni decided against joining the WMA was ward-level politics. The Makuyuni ward councilor had been a strong advocate for the WMA, knowing that it could be a way to raise money for the district. While the hunting block in the Game Controlled Area (GCA) generated considerable revenue, those funds were mostly taken by TAWA. The WMA was seen as a way to potentially tap into the profitable agreement between Treetops and Lolkisale and ensure that some money was also coming back to the district. The community of Makuyuni was well aware that their ward councilor supported the WMA and came to associate the WMA with the councilor, believing the WMA proposal to be part of his personal vision and agenda. This association was manipulated by other political hopefuls who opposed the councilor and were aspiring to succeed him in the subsequent election. Several candidates had made visits to Makuyuni as part of their campaigning efforts, where they used polemical arguments to invalidate the planned initiatives of the councilor. Many community members in Makuyuni were swayed by these visits and began to perceive the councilor with skepticism. Consequently, they also began to view the proposed WMA with distrust because of their conflation of the councilor's political standing and the prospect of a WMA. In actuality, these sentiments were entirely political and had little to do with the material aspects of the WMA.

Ultimately, this combination of ethnographic factors led the community of Makuyuni to withdraw from the WMA proposal. During interviews with Makuyuni residents in 2019–2020, some people alleged that because Makuyuni had dropped out of the plan late, a portion of its land had already been included in the WMA's reserved area. The WMA indeed came to include MEDA, and Makuyuni was never compensated for this contribution. Following Makuyuni's decision to withdraw from the WMA in 2011, the proposal was reduced to four villages, though Lolkisale was preparing to subdivide at that time. Unlike Makuyuni, the other additional villages—Naitolia, Mswakini Juu, and Mswakini Chini—ended up joining the WMA, but their processes for taking this decision were highly contentious. After their representatives returned from the scouting trip to Enduimet WMA, the villages of Naitolia and Mswakini (Juu and Chini) came under significant pressure from AWF and different levels of government to accept the WMA. Unlike Makuyuni and Lolkisale, there was a lack of governance transparency in these villages at the local level, and the decision to move forward with the WMA was taken without input from the majority of the village assemblies. These gaps in participation led to the corrosion of community support for the WMA in the early stages of its establishment (Loveless 2014).

Though the idea for a community-based conservation area had developed organically in Lolkisale village, part of the motivation for reforming the LCA into a WMA had come from the central government. Around the same time that the Kisongo-led party from Lengoolwa carried out its scoping mission to Longido, the central government called a meeting in the country's capital, Dodoma, to articulate the Ministry of Natural Resources and Tourism's (MNRT) commitment to the WMA model of conservation on village land following the revised Wildlife Conservation Act of 2009. During the meeting, ward councilors from Mswakini, Makuyuni, and Lolkisale were directed by representatives from the MNRT to facilitate the implementation of a WMA in their jurisdictions (Loveless 2014). The councilors were explicitly asked to persuade village leaders to join the WMA. As discussed in the previous section, the assembly of Makuyuni was wary of the Makuyuni ward councilor and, due to the village's strong governance institutions, made a collective decision to withdraw from the WMA. By contrast, Naitolia, Mswakini Juu, and Mswakini Chini bent under the influence of the central government, as intended by MNRT. The ministry had hoped, however, that the village leaders would internalize the suggestions of ward councilors and subsequently teach their councils and development committees about the benefits of establishing the WMA. The ministry expected that knowledge about the WMA would then trickle down from the level of village councils to the general assembly, thus manufacturing community support from the top down (Loveless 2014). Monduli District Government and AWF, the latter of which served as an "NGO facilitator" throughout the process, were tasked with helping to fast track this top-down process of WMA implementation from the central government down to the community level (Loveless 2014).

Sensitizing Naitolia and Mswakini

In 2011, the central government and AWF undertook a process of "sensitization" involving outreach work at the village level (Loveless 2014, 45). Sensitization was meant to soften the villages of Naitolia and Mswakini to the idea of establishing a WMA. In accordance with ministry guidelines, village councils were targeted as potential liaisons between the central government and village assemblies and were expected to educate people about the WMA and advocate for community concerns. In practice, however, they were only accountable in an upward fashion to the government, and village-level education initiatives were ineffectively implemented. Thus, sensitization did not take hold.

While consultation with communities technically featured in the ministry's prescribed WMA procedures, little effort was made to meaningfully engage the Arusha living in Mswakini and Naitolia. Unlike the democratic meetings in Lolkisale and Makuyuni, the majority of villagers in Naitolia and Mswakini were not represented in WMA-related assemblies. Two main planning gatherings were held in Naitolia and Mswakini (Chini and Juu) villages in 2011. In the context of an anthropological Master's study, Loveless (2014) documented attendance rates of less than 10 percent of the total village populations and noted that the meetings were structured in ways that foreclosed meaningful discussion. Allegedly, village leaders even exaggerated attendance rates and forged lists altogether. Community members further accused that key committee members had hidden details of WMA meetings from the assembly to concentrate decision-making power in the hands of a select few. During the meetings, sensitization teams gave cookie-cutter responses to general questions but did not engage the assemblies in meaningful discourse. Community involvement in the WMA planning process was thus largely tokenistic in Naitolia and Mswakini—a box to tick before moving on to the next stage of implementation. Importantly, this was very different from what had transpired in Makuyuni and Lolkisale, where there was considerable discussion and debate within the communities about the benefits and drawbacks of establishing a WMA. Low attendance rates at the WMA meetings in Naitolia and Mswakini meant that most people in those villages only heard about the WMA "through the grapevine" during their everyday conversations with friends and family members. Very few people in those villages were properly informed about the WMA or aware of how it would work. The cracks in communication between village government and the assembly meant that misinformation about the WMA and distrust were rampant. Villagers suspected that their local governments had vested interests in supporting the WMA, undermining the ministry's attempts to sensitize the communities about WMA benefits (Loveless 2014).

Exclusionary local governance institutions meant that very few people in Naitolia and Mswakini were able to track how the WMA planning processes were progressing. As Loveless (2014) found through surveys in those villages, approximately half of her respondents were unaware of the WMA. Her research was carried out two years after the WMA had been officially gazetted, indicating that community outreach was still minimal after initial implementation. Disturbingly, she also found that the majority of people she surveyed had concerns about the WMA but had not expressed them to their leaders. This was particularly so for women and youth, whose voices were given less authority in governance processes. Some of these gaps in awareness have persisted. I found during my fieldwork in Naitolia and Mswakini in 2019–2020 that some people only had general ideas of what the WMA entailed. People generally knew enough to answer basic survey questions about them, but when they were prodded on the more complex characteristics of WMA governance and management, some did not fully grasp what it actually involved. Notably, this was not the case on the Lolkisale side of the WMA, where most people I interviewed had a solid understanding of what the WMA was and how it operated. This trend was more apparent among women in Mswakini and Naitolia, some of whom expressed less nuanced opinions on the WMA during interviews than male heads of houses. This knowledge gap was perhaps related to the traditional gender-based divisions of labor in Maasai society, as men generally participate in governance meetings where land and grazing issues are discussed, while women tend to households and rear children. Notably, however, some anthropologists have critiqued the assumption that Maasai women are excluded from governance meetings altogether (Hodgson 2001). What is certainly clear is that the entire sensitization process in Mswakini and Naitolia fell short due to poor local governance. As a result, many community members, and especially women, were unaware of what the central aims of the WMA were. The lack of education about the WMA created a barrier to genuine participation in the planning processes and foregrounded growing discontent within those villages about the WMA's establishment.

Ultimately, the village councils of Naitolia and Mswakini agreed to implement the WMA in their villages without adequate input from their constituents. While these leaders were guilty of poor governance practices that limited participation from their assemblies, they likely also faced coercion from above. Some village committee members felt that opposing the WMA would be futile because there were larger forces at play. When the village chair of Naitolia vocalized opposition to the WMA, he was excluded from further WMA planning and governance meetings from then on. Indeed, there were big players behind the scenes pushing the WMA through, some of whom Loveless's (2014) interlocutors felt insecure even mentioning.

The major politician involved was Edward Lowassa, who at the time served as the Member of Parliament (MP) of Monduli before leaving that position to become the prime minister of Tanzania under the ruling party, Chama Cha Mapinduzi ("party of the revolution"). Lowassa would later defect to Chadema, the leading opposition party, after he was not selected as Chama Cha Mapinduzi's presidential candidate. He ran against the late President Magufuli in the 2015 presidential campaign and narrowly missed winning. I had a chance to chat with the former prime minister at the home of a mutual friend in Monduli town in early 2020, though I did not formally interview him about the WMA. In Tanzania, MPs have the authority to debate bills and pass laws and consequently have much power in the policymaking arena. Lowassa was supplanted by Julius Kalanga as the MP of Monduli from 2015–2020, and I interviewed Mr. Kalanga about the WMA in mid-2020. While Lowassa served as the MP of Monduli, he went to considerable lengths to push the WMA through to formalization as quickly as possible. I have heard several different perspectives on why Lowassa was so interested in fast-tracking the WMA. The first was that he appreciated keeping livestock as a way of storing and accumulating wealth outside central banks and viewed the rangelands of Monduli as an appropriate place for managing herds. This aspect of his policy was certainly favored by the Kisongo pastoralists living in the Lolkisale area. Some of my key informants from district government suggested to me during interviews that Lowassa was genuinely interested in contributing positively to pastoral development. These same interlocutors relayed to me that Lowassa had been troubled by the alienation of pastoral land for commercial farms, agricultural expansion, and national parks, and saw the WMA as a means of defending the land rights of local pastoralists. At the same time, some of my local Kisongo interviewees resented him because they felt that he was looking for areas to privately manage his own large herds without heed to customary Maasai range management institutions. Several of my interlocutors suggested to me that Lowassa's plan was to keep his own herds in the WMA area, and this was the main reason for his interest in encouraging its establishment. Other key informants at the district level, however, were adamant that there was no evidence to suggest that those allegations were anything more than rumours. Another explanation raised by some members of district government was that Lowassa saw the WMA as a legitimate strategy for ensuring that revenue from community-based tourism in Lolkisale would make it back to the Monduli District Council, rather than remaining in the villages or being "eaten" by TAWA.

Another key government figure who supported formalization of the WMA was the Monduli District Game Officer, Seraphino Mawanja. I was able to formally interview Mr. Mawanja in 2020 to garner his views on the WMA. I later crossed paths with him several times in Nafco village and at a number of WMA governance meetings in Makuyuni, Mto wa Mbu, and Monduli town. Mr. Mawanja was not acting out of personal interest in helping to establish the WMA, though he was certainly not opposed to generating more money for his branch at the district government, which was generally constrained by a lack of funds. Mawanja is a dedicated conservationist who genuinely cares about the well-being of wildlife and the equitable distribution of conservation benefits to communities. Prior to being placed in Monduli District, he worked in Morogoro and advocated to TAWA that a major trophy hunting block be closed because he felt it was significantly impacting wildlife populations. TAWA abruptly relocated him to Monduli District where he had since been working for many years at the time of writing. Mawanja's interests were closely aligned with the Kisongo of Lolkisale: he wanted the GCA to be reclassified as a photographic tourism area because he felt that would be of greater benefit to wildlife and local communities. When I first met him at a WMA governance meeting in early 2020, he was outspoken about the need for a resolution of the ongoing Boundary Hill court case and felt strongly that Boundary Hill

Lodge should pay Lolkisale village back what it was owed. In May 2020, when I was staying in Nafco village, Mr. Mawanja appeared on his day off with a district vehicle to help local Arusha cultivators herd elephants back into the WMA. The community very much appreciated these efforts. Mawanja prides himself on having boots on the ground and considers himself to be a rogue actor who follows his heart irrespective of what the central government expects of him. He is an endearing character, and I think highly of him. When I explained the nature of my research and my interest in evaluating community perspectives on the WMA, he very thoughtfully explained that I must make sure to attend to issues of class, education, gender, and geography because the opinions of people might vary significantly and that I "should make sure to get all the opinions of people to really understand what is going on." Mawanja is also a strong-minded and charismatic fellow, and I suspect that he was fairly direct in his suggestions to members of local government that they should support the WMA, though I believe his reasons for advocating for the WMA were well meaning.

Regardless of the intentions of Lowassa, Mawanja, and the three ward councilors, Loveless (2014) noted that several members of village government in Naitolia and Mswakini felt strong-armed into accepting the WMA. Her interlocutors suggested that money was dangled as an incentive to get village governments on board. MP Lowassa advised ward councilors and village committees to move swiftly to tap into the revenue flows from community-based ecotourism. Central government officials also promised infrastructure development in the villages and pointed to the schools and dispensaries in Lolkisale village as examples of what to expect. Perhaps most worryingly, Loveless's (2014) informants reported being convinced through threats of displacement, which motivated village governance committees to take acceptance decisions prematurely. On the Lolkisale side, there were indeed concerns that if the WMA was not established, village land could be reclassified as part of the GCA, potentially undermining people's rights to graze livestock and collect firewood, not to mention Lolkisale's partnership with Treetops. In Mswakini Juu, where conflicts over the boundaries of Tarangire NP and village land were ongoing, one member of village government was told that if they did not accept the WMA in their village, "Tarangire would extend its borders and they would remain with no land at all" (Loveless 2014, 56). A wildlife corridor was also floated as an alternative possibility in Naitolia and Mswakini that would evict the villages entirely (Goldman 2009). According to one member of the Naitolia council, MP Lowassa had advised him directly that the WMA was the only option available to prevent the central government or foreign investors from grabbing the land (Loveless 2014). If he did not accept the WMA, then the area would be converted into a corridor. The local governments of Naitolia and Mswakini were thus motivated to join the WMA despite its potential risks because it at least ensured that they would not be displaced altogether by a national park or wildlife corridor. The WMA was seen as a lesser of two evils.

The Arusha were particularly cognizant of being evicted, knowing that they were relatively recent migrants and that they had moved into an ecologically significant wildlife dispersal area. The Arusha-led village councils in Mswakini and Naitolia were assured by ward councilors that the WMA would uphold the land rights of Arusha cultivators and strengthen their tenure security in the face of uncertainty. The village assemblies, however, were not convinced by this claim. As one local government member in Mswakini Juu explained to Loveless (2014), people in his village were anxious about establishing a WMA, knowing that Mswakini and Naitolia were already surrounded by wildlife areas where villagers did not have resource rights. Village governance committee members also held this fear but channeled it into support for the WMA, which they viewed as a means of securing land. By contrast, the village assemblies worried that their local governments had signed off on the very thing that people were afraid of—a wildlife conservation area that would displace them from their homes. Community members were concerned that, given all the other protected areas around, this would turn into yet another situation where the government sought to extract tourism revenue from wildlife at the expense of local livelihoods. Some feared that the government was tricking people into thinking it was a community-based idea, when in reality, it would be used as a precursor for eviction (Loveless 2014). People had seen conflicts arising in Burunge WMA over livestock grazing rights, and they were also concerned that the WMA might implement a multiple land use model like the Ngorongoro Conservation Area that prohibited cultivation (Loveless 2014). This was a major concern for the Arusha since they were mainly farmers.

In Naitolia, environmental and development committees were at times at odds with the village councils, as they had been directly influenced by ward councilors and district government officials. The former village chair of Mswakini Chini explained to Loveless (2014) that when Monduli district government officials communicated the costs and benefits of the WMA to the village committees, they overemphasized the benefits (income gains, employment opportunities) and underemphasized costs (increasing human-wildlife conflict, loss of land). As a result, some members of the village committees had come to expect that the WMA would generate unrealistically high revenue for the villages relative to the costs and benefit-sharing structure of the future WMA (Loveless 2014). Furthermore, district officials underreported the impacts the WMA could have on local livelihoods. Men in Naitolia and Mswakini were particularly anxious about their farms and livestock, while women were worried about restrictions on collecting firewood and building materials (Loveless 2014). These concerns, however, were neglected in the decision to establish the WMA and in the process of devising its initial management plan.

Establishing the WMA

The process of formally establishing the WMA followed national guidelines and a 2011 handbook from AWF. The WMA was named Randilen (formerly *Randileni*), the Maasai place-name given to the hill (after a man named *Sandilen*) near the proposed WMA's entrance gate, which marked the border between the preexisting LCA and Naitolia Camp (King 2009). Randilen had been proposed as an entrance gate and visitor center as part of the original Tarangire Conservation Area (King 2009). The first step in the formalization process in 2011 was establishing a community-based organization (CBO), a body of representatives from each member village that would later become an Authorized Association (AA) once the WMA had been made official by the ministry. The AA would then be responsible for making governance decisions affecting the management priorities of the WMA. Selection of CBO candidates was supposed to be based on transparent elections, but in the case of Mswakini and Naitolia, individuals with other personal leadership positions were tapped to become voting members of the CBO. The preselection of CBO candidates was partially about consolidating power within the leadership groups of these villages, but it was also necessary because there were simply not enough nominations received from the general assemblies (Loveless 2014). Mswakini Juu, for instance, only received two applications for the position, and the rest were handpicked by the environmental and development committees (Loveless 2014). As such, governance authority was assigned to preselected villagers and not fully devolved to the level of the assembly. By contrast, villages on the Lolkisale side selected CBO representatives democratically through good governance institutions at the local level.

Once members from each village had been selected, the CBO became tasked with preparing the WMA's constitution, strategic vision, and resource zoning management plan (RZMP). The zoning plan had to be prepared using official ministry templates, and many of the community members did not have the technical expertise to know how to work on them (Loveless 2014). AWF played a significant role at this point by helping fill out the paperwork and write up the resource use plan. Once the documents were finalized, the CBO then applied to the ministry for AA status and formally became an official governance body in the eyes of the state in 2012. This meant that the AA superseded the governance authority of each individual village in the context of administering WMA resources. The WMA was gazetted in accordance with TAWA's new set of official WMA regulations that were updated that same year. Following the establishment of the AA, the Director of Wildlife granted the AA wildlife user rights, which bestowed upon it the authority to make governance decisions about the WMA and negotiate contracts with investors. A bid and tender committee was established, including members of the ministry, Monduli district government, and the AA to formalize contracts with investors, though final authority for approving deals rested with TAWA (Loveless 2014). Importantly, preexisting direct photographic tourism investments in villages, like those seen in Lolkisale, were no longer permissible under state law, so all future negotiations had to be carried out with the AA rather than village councils (see also Kimario et al. 2020).

Unique to Randilen WMA compared to other WMAs was that the Tarangire Conservation Area already had a fairly comprehensive eighty-page management plan that the Australian investor had revised in 2009 using the development loan. Thus, devising Randilen's initial management plan was not as difficult as it had been with some other WMAs because the planners did not have to start from scratch. Some of the proposed features of the WMA, like the location of the entrance gate and ranger posts, remained from previous iterations of the plan. This jumpstart helped get Randilen off the ground quickly. In fact, it was one of the fastest WMAs ever formalized in Tanzania. From the time of the initial planning meetings in 2011 to when it was officially gazetted in 2012, it took around a year. Its rapid establishment was due in large part to the political support from above, as MP Lowassa was intent on pushing the WMA through. It also would not have been possible without the technical assistance of AWF, particularly in the preparation of the initial zoning plan. Village governments in Mswakini and Naitolia reported feeling considerable pressure from AWF to finalize the necessary documents and paperwork (Loveless 2014).

On the Lolkisale side of the WMA, however, the village council and AWF collaborated closely. Making the RZMP required the use of Geographic Information System technology to map out the area and its boundaries. Local herders did not have these technical skills, nor the equipment to gather GPS points, analyze them on a computer, and prepare maps. AWF filled this technical gap by preparing maps for the initial RZMP in collaboration with Monduli district government. However, key issues emerged during this process. From the perspectives of community members, AWF had been in such a rush to carry out the mapping that they did not do a thorough job of assessing where the boundaries ought to be. They relied on the directives of a few individuals rather than the community at large. Consequently, the initial maps were not representative of local livelihoods concerns. In Lemooti, for instance, the Kisongo were angry about the way the map of their village had included a key permanent water source inside the reserved area (D. Bell, personal communication, 2019). From the perspective of the mappers, the difference was miniscule, as it was only a matter of a few hundred meters. From the perspective of local Kisongo, however, it was a crucial difference. After a follow-up to inquire why the water source had been included in the reserve area even though the community had not wanted it to be, the Monduli district government official admitted his error. He explained that the boundaries on the ground were likely not exact because he had been under pressure from AWF to formalize the maps as quickly as he could. AWF at the time was working toward achieving key deliverables for their donors, measured in terms of how many WMAs they had established across the heartland of Maasailand. They were thus driven to establish as many as they could in a short period of time. In practice, this led to mapping issues that were very significant from the perspectives of local community members.

Some of the boundary disputes were between villages, which would have been an issue irrespective of the WMA. Mswakini Chini and Olasiti village (in Babati district), for instance, had an ongoing dispute over village boundaries, made even more complicated by the fact that they were located in different administrative regions (Arusha and Manyara) (RWMA 2018). Makuyuni also had a conflict over its village boundaries that remained unresolved at the time of writing. Makuyuni's issues had been due in part to political complexities stemming from the cancellation of Steyn's lease and the limbo status of some of the pastoral areas adjacent to Saburi near Esimangore mountain. As described to me by the urban planner of the Monduli Government, Makuyuni's boundaries had still not been formally mapped in 2020 due to the unresolved disputes. Some of the villagers I spoke with in Makuyuni in 2019 felt that the lack of resolution of the formal boundaries of Makuyuni was a contributing factor that allowed Randilen WMA to include a portion of land in its reserve area that would have otherwise technically been in Makuyuni. In 2023, the state finally clarified the situation in a stronghanded manner by declaring the remaining portion of Steyn's former farm as "Makuyuni Wildlife Park," a TAWA-run protected area that prohibits local livestock grazing.

The assemblies of Naitolia and Mswakini Juu were also concerned about how far the WMA's reserve area extended onto village land and were unclear about where the exact boundaries were on the ground. Some people believed that the WMA's border between Lemooti and Naitolia passed through the Santilen hills (later renamed "Randilen"), but the RZMP indicated that the boundary moved in a different direction near the hills (RWMA 2018). Community members in Mswakini Chini argued that the visual landmarks they were accustomed to using to demarcate boundaries were not represented by the management plan (RWMA 2018). Some of these boundary conflicts had led to increasing tensions, particularly in Mswakini Juu and Naitolia.

While committees were formed to prepare the initial zoning and land use plans, they were led by technical experts and included only a few village representatives through the CBO. Rather than provide community members with opportunities to contribute to the formulation of the plans, the committees prepared them independently before presenting them to the villages afterward (Loveless 2014). As such, the initial WMA zoning plan did not represent the voices of the assemblies in Mswakini and Naitolia. Technical knowledge was prioritized through the influence of AWF, and community livelihood concerns and local ecological knowledge were underrepresented (Loveless 2014). This led to disillusionment when the initial management plan was put into practice in 2014.

Though the WMA was technically gazetted in 2012, it was not yet operational until April 2014 when village game scouts began implementing beacons and patrolling the area. It was at that point that residents of Naitolia and Mswakini were finally informed by AA members and district government officials about local land use restrictions during village assembly meetings. Most community members, however, did not attend those meetings and were still unaware of the WMA and how it functioned. They also did not know who the AA members were and what exactly they represented (Loveless 2014). People were confused when they encountered game scouts and beacons in their villages and in some cases had no idea why they were even there (Loveless 2014). They grew scared that the WMA was going to be used to "police the community" and became deeply resentful of the exclusionary way it had been designed (Loveless 2014). Despite having little say in the determination of the WMA's management priorities, local Arusha knew that they would be the ones directly affected by the new regulations. Understandably, they were afraid, confused, and distrustful of the changing political landscapes of their villages. People felt as though their village leaders had secretly sold their land in cahoots with the central government, dispossessing the community of its land. As one sub-village chair in Mswkaini Chini explained to me during an interview in 2019:

> From our perspective in Mswakini, the idea of a WMA started from the government. They came to the village offices and said that local government members should go and spread the idea of having the WMA. People protested here because they thought that the land had been sold. We thought that we would no longer be allowed to access the forest, so we were very worried when it first started. The WMA did not start in a good way because it did not educate people about what the WMA was and about its benefits. There was a lot of misunderstanding at that time and this led to conflicts between the WMA and the people of Naitolia and Mswakini.

Tensions began to rise in Naitolia and Mswakini when one man was caught in the reserved area of the WMA on a motorbike and was reportedly beaten by game scouts, though he was apparently unaware that he had transgressed any boundaries (Loveless 2014). Some community members alleged that seasonal *ronjo* bomas (short-term settlements for mobile herders) in key grazing areas had been destroyed in accordance with the new zoning plan. Herders interpreted these changes as top-down exercises of power over community land and livelihoods.

Enforcement of the WMA's initial zoning scheme greatly angered local livestock keepers and, in 2014, a group of concerned citizens from Naitolia and Mswakini stormed the village offices and threatened WMA supporters with violence. Those targeted locked themselves inside small rooms until the angry mob passed. One individual who experienced the confrontation told me during an interview that if they had not done so, they might have been killed. As community discontent took hold, residents of Naitolia and Mswakini rose up in opposition to the WMA, and the incident escalated to a full-scale blockade of the A104 Arusha-Babati highway in mid-2014. People carried signs, yelled, and stopped cars by marching across the highway in protest. They were led by the village chairs from Naitolia and Mswakini who had expressed concerns throughout the planning processes but whose voices had been overruled, first by the development committees and later by the AA (Loveless 2014). The protests were a clear sign that the Arusha would not willingly stand by while their land was being grabbed. As described during an interview with the head of the rangeland committee in Mswakini Juu in 2019:

> Before the WMA started, the whole area was our livestock grazing area. The only place where we were not going was Tarangire, but this was our land. We were grazing there. When they came and introduced the idea of the WMA, a lot of people thought that this land was being taken away. So, when the WMA implemented signs and beacons to mark its boundaries, we went and destroyed them. We were afraid of the WMA at first. We did not have enough education about what was going to go on within the WMA. We had no concept of what a WMA was since people here only knew about national parks.

The Arusha of Naitolia and Mswakini were concerned that they would be subjected to new rules affecting their livelihoods that they had never agreed to in the first place. These included restrictions on firewood collection, charcoal production, livestock grazing, and farming. As a substantial body of literature shows, rural communities do not roll over and accept the fates that are prescribed to them by governments through top-down conservation (Bluwstein and Lund 2018). Rather, they often exercise their agency by resisting these political forces (Holmes 2007; Raycraft 2016, 2019, 2020). Residents of Naitolia and Mswakini were not passive victims but engaged agents who mobilized in defense of their land and livelihoods. To Loveless (2014), the acts of protest were direct results of top-down governance practices and disregard of community concerns during the creation of the WMA. As Brockington and Igoe (2006, 2) point out, protest against conservation "is likely to be loud where people are highly dependent on natural resources for their livelihoods and risk facing impoverishment because of those regulations."

In response to the agitations, the state swiftly intervened by sending in paramilitary forces and police officers to break up the crowds and arrest the village chairs, who they targeted for mobilizing the communities. The chairs were subsequently imprisoned for two weeks, and the village offices in Naitolia and Mswakini were closed while the dust settled and the situation de-escalated. The WMA, however, had already been formally gazetted, and the boundaries remained on the ground. As described to me during an interview with the WMA manager in 2020, people in those villages still opposed the WMA throughout the rest of 2014 but were afraid of voicing their discontent after the state's demonstration of power and authority. People instead resorted to "everyday resistance" in the form of Arusha curses that were cast with the intention of destabilizing the WMA and avoiding open conflict with governing authorities (Hoffman 2014; Holmes 2007; Scott 1985).

Commercial Farmers and Cattle Barons

Loveless's (2014) study shows the effects that poor local governance and inadequate community participation in WMA planning had on people's conservation attitudes in Naitolia and Mswakini when the WMA was first established. While the protests were certainly a reflection of people's concerns about land tenure and livelihoods, my interviews with government officials, WMA management, and several nongovernmental organization (NGO) representatives also point to another set of political factors at play that does not feature in her work—the role of a minority of powerful elites who viewed the WMA as a threat to their private interests. While Loveless (2014, 58) rightfully suggests that local Arusha were "duped into accepting" the WMA, here I would add that they were also subsequently duped into opposing it. As Brockington and Igoe (2006, 2) also recognize, "protest is likely to be loud when those affected are wealthy and powerful and not able to become richer and more powerful as a result of the restrictions of conservation." Behind the scenes, open opposition was driven by a strong and vocal minority of domestic elites who were able to manipulate Arusha villagers through fear mongering. While there were valid reasons for the Arusha in Naitolia and Mswakini to oppose the WMA, influential elites also played a significant role in swaying those villages toward protest.

Absentee commercial farmers in Mswakini and Naitolia saw the WMA as a direct threat to their operations and had no interest in having their farms repurposed as a reserve area. As articulated by an Arusha man (about fifty-four years old) from Naitolia during an interview in 2019, "Before the WMA was established, there were some people here who had grabbed large plots of land for commercial farms. When the WMA was created, those people who knew that their land was going to be taken were angry so they came and spread the idea that the WMA would be a park that evicts the whole village." While the commercial farmers in Lemooti had capitalized on the land scramble following the cancelation of Steyn's lease, those in Naitolia and Mswakini had secured land through their dealings with the village councils, which constituted formal process under state law pursuant to the Village Land Act. Absentee farmers were prominent in Naitolia and Mswakini, where wealthy Meru, Arusha, and Chagga who had made significant profits from the commercial coffee trade in Meru sought to reinvest in farmland in rural Monduli. They picked areas where land was affordable (away from the major city), where kinship ties could be leveraged, and where village councils were "agreeable" to allocating village land to wealthy individuals from outside the villages. Additionally, some outside farmers had purchased villagers' Certificates of Customary Rights of

Occupancy, which were usually signified by word-of-mouth agreements or a letter from the village government. One of the key strategies employed in those villages was accumulating and trading small parcels of farmland allocated to individuals to consolidate a large area that could be cleared for a commercial farm. One Chagga commercial farmer I interviewed in Makuyuni town in 2019, for instance, described to me how he had purchased ten acres here, and another five there and bought and traded his way to a two-hundred-acre farm in Naitolia. During interviews with Arusha community members in Naitolia, people consistently expressed suspicions of corrupt land dealings within the village. Several key informants accused the former village chair of being easily convinced with gifts, alcohol, and money to allocate farmland to outsiders and favor their claims over local Arusha residents who had inherited farms from their fathers. A key issue was that land acquisitions within villages often took place without formal paperwork. This made it challenging for those with less political power to stake their claims and straightforward for those with "money power" to bully their way into the villages.

Further to commercial farmers, elite livestock keepers were also involved. A regional politician had allegedly paid local herders to take care of his large herds and negotiated access to communal grazing areas through the village council of Naitolia. The former chair of Naitolia, who Loveless (2014) noted was opposed to the WMA, had made private arrangements with this politician to allow him grazing access in Naitolia village. The issue was made even more complex by class differences among Kisongo livestock keepers, some of whom had accumulated herds of more than a thousand cattle. The emergence of Kisongo elites over the past thirty years in the Lolkisale area is attributable in part to the commodification and globalization of the tanzanite industry. Not far from Lolkisale are the Mererani mines, which constitute the only place in the world where one can extract tanzanite, a precious gemstone of global significance. While Merarani is technically located in the Simanjiro District of the Manyara Region, which makes up Tanzania's southern Maasailand, the Kisongo of Lolkisale are related through kinship ties to the Kisongo of Simanjiro. With the emergence of tanzanite as a globally coveted gemstone, artisanal mining in Mererani became a ticket for many Kisongo to a life of wealth. Despite tightening grips of state and corporate interests over the mines in recent years through privately leased commercial blocks, the artisanal mining segment of Mererani still offer opportunities for lucrative payouts for skilled, dedicated, and downright lucky artisanal Kisongo miners. A general distrust for central banks, and an awareness that cash faces a losing battle against inflation in the long haul, has led many of these new elites to reinvest the profits in livestock. This was the approach taken up by one young man from Lemooti who managed to amass a sizeable fortune from mining and trading tanzanite. Though he was from Lemooti sub-village, he lived primarily in Arusha town and traveled back and forth from the mines to trade in stones. He had managed to amass a herd of around two thousand cattle thanks to his continual reinvestments, which he divided across Simanjiro and Lolkisale. When the prospect of the WMA was raised, he was particularly opposed to it, knowing that it might jeopardize his private interests in the area. Realizing that he was outnumbered on the Lolkisale side by those who supported the idea of establishing the WMA, he allegedly began spreading rumours about the WMA across Naitolia and Mswakini to generate community resistance. As described by an Arusha man (about fifty-four years old) in Naitolia during an interview in 2019:

> There were five rich herders from outside our village that were spreading all the bad ideas about the WMA. They wanted to use the land for their own benefit because they had large herds. You know, that place was like a forest without much government monitoring. They wanted to put their own bomas in that area and keep it for their own livestock. Before the WMA came, those rich herders were trying to remove us community members from that area so they could keep it for themselves. Our community quarreled with those people for ten years before the WMA came. But when the WMA was established, it helped us push those people out of our community.

During several of my interviews with key stakeholders with knowledge of the initial implementation phase of the WMA, it was brought up that some of the "cattle barons" (as the rich herders were referred to by some interviewees) and commercial farmers had taken measures to destabilize the process of formalizing the WMA. The village chairs that Loveless (2014) acknowledged as being opposed to the WMA, for instance, were being lobbied by the politician and commercial farmers who had made private arrangements to access farmland and pasture in their villages. When the village chairs were thrown in jail during the protest, the police knew about their connections to those figures and advised the chairs to cut ties with them. The outside elites were well aware that the WMA would undermine their access to village land and used their money and political influence to sway the general populations of Naitolia and Mswakini to revolt against the WMA. As one representative of AWF explained to me in an interview in 2020, some of the politicians involved even hired journalists and news reporters to spread fear in the communities and stoke the flames of discontent by broadcasting the rumor that the local villages in the area were going to be evicted to make way for a new national park.

While the Lolkisale community was generally aware of what the WMA would entail, the Arusha-inhabited villages of Naitolia, Mswakini Chini, and Mswakini Juu had largely been kept in the dark and were very fearful that they were going to be relocated. The elites who had existing stakes in those villages manipulated and stoked people's fears by encouraging the village chairs to incite opposition at the level of their assemblies. During an interview with one elder Arusha woman (about seventy years old) in Naitolia in 2019, I asked if she remembered a time when people opposed the WMA. She replied:

> Yes, there was a big protest. We followed people there, but we did not know what had happened. I went with my sister and someone gave us a sign. We were all screaming and yelling and marching and then the police came and shot their guns. Some people fainted so we thought they had been killed! So my sister and I looked at each other and asked "Why are we even here? Why did we follow these people? What is this protest really even about anyway? Are we willing to die here?" We dropped the sign and ran back to our houses and hid inside! *starts laughing*

As the woman's narrative segment reveals, in the same way that people were uninformed about what the WMA actually was, some were also unaware of what the reasons were for opposing it. On the surface, it might have seemed to Loveless (2014) like the entire villages of Naitolia and Mswakini were opposed to the WMA, but there was much misinformation going around, and there were only a few key players who were fully informed about the WMA and the protests against it. Ordinary residents of Naitolia and Mswakini were caught up in the crossfire of a larger conflict between powerful political actors. On the one side of the struggle, MP Lowassa, Monduli DGO Mawanja, the Mswakini ward councilor, and AWF staff were pushing swiftly for the villages to accept the WMA without adequate deliberation and discourse; in the other corner, the cattle barons and commercial farmers who already used the area for their own personal gains, and who had much to lose by the prospect of tighter environmental regulations, stoked the flames of discontent. Local Arusha were subject to manipulation largely because of the lack of awareness and participation that Loveless (2014) documented during the initial stages of WMA planning. People did not have concrete facts about what the WMA would entail in practice, and this lack of knowledge allowed rumors to flourish. The Arusha were largely disregarded by the powerful actors who wanted to establish the WMA and used as pawns by the elites who opposed it. All the while, people felt genuine and well-grounded apprehension about what the future would hold for their land tenure rights and livelihoods.

CHAPTER FIVE

The Rise of Randilen

When I commenced fieldwork in Naitolia and Mswakini in July 2019, I initially expected to find widespread community-level opposition to the Wildlife Management Area (WMA), as had been documented by Loveless (2014). What I found in practice, however, was that residents of those villages cherished "their" WMA and its crucial role in securing their livelihoods and land. Through interviews and informal conversations across the member villages, this sentiment continually shone through. People explained that they were "proud of their WMA" and spoke with sincere affection about its positive impact on their everyday lives. I became determined to assess the generalizability of my findings across the whole community, which led me to design and implement a quantitative survey at the household level to assess, in a representative fashion, people's general sentiment toward the WMA across the entire population (see introduction). I used systematic random sampling, with sub-villages as strata, to survey a representative sample ($n = 678$) of male and female household heads across Randilen's eight member villages (Raycraft 2022a, 2022b). Rather than stand alone, the survey results triangulated my ethnographic findings and qualitative data by providing a "quantitative snapshot" of community sentiment toward Randilen WMA. I can assert with confidence that the findings of the survey did indeed mirror my qualitative findings, elicited through hundreds of in-depth interviews and participant observation of everyday life while residing in Randilen's member villages.

That said, the survey results should also be interpreted with a grain of salt, and there are a few disclaimers that are worth mentioning as a preface. First, while respondents were assured of anonymity during the conduct of the surveys, names were randomly selected from lists of households and used to follow up with respondents if there were inputting errors or missing responses. Respondents may have had some fear that voicing opposition to the WMA could be construed as critique of government, which community members were apprehensive about during Magufuli's second term as president. Second, there seemed to be "research fatigue" in the villages near the main highway (Naitolia and Mswakini), and it is possible that people felt inclined to simply answer the survey questions in the way they viewed as the path of least resistance. Some of these constraints on survey research are inevitable.

Bearing in mind the potential limitations of the survey method I used, there were also some factors that speak to its validity: respondents were randomly selected, and the samples were representative and proportionately weighted across sub-villages relative to population sizes of each village. Seven different field assistants helped administer the surveys, each of whom documented similar findings independently. The survey results also aligned with my observations, informal conversations, and interviews with community members, giving me greater confidence in interpreting the results.

Overall, the survey results were remarkable. Three-quarters of respondents ($n=508$) across all eight villages liked the WMA, and about 76% ($n=513$) supported its presence in their villages (see figure 1). Almost 60 percent of people ($n=397$) reported feeling more positive toward the WMA now than they did five years ago. Rather strikingly, 88 percent ($n=594$) of participants trusted WMA authorities to act in their community's interest. When asked about conservation trade-offs, and people's perceptions of the distributions of costs and benefits associated with the WMA, three-quarters ($n=511$) of respondents reported that it had more benefits than costs. Perhaps most significantly, 92 percent ($n=624$) of respondents felt that their community was well represented by WMA governance, and 91 percent ($n=620$) thought that their community was included in WMA management. To sum up their perspectives on Randilen, respondents were asked whether they viewed the WMA as a success or a failure, and whether they felt it represented a top-down strategy for securing resource control at the expense of local communities (i.e., "fortress" conservation) or a community-based conservation area that served local interests. Results showed that 90 percent ($n=607$) of respondents felt that Randilen WMA was a community-based conservation area rather than a fortress model, and 94 percent ($n=634$) viewed it as a success rather than a failure.

The quantitative results do not tell the whole story, but with a grasp of the ethnographic context, they tell a fair amount. There are a few culturally specific details worth noting. The neutral category was translated on the written survey as *kawaida*. Though *kawaida* is a kiSwahili word meaning "normal" or "usual," it is also used by the Maasai. *Engishui e kawaida*, for instance, means

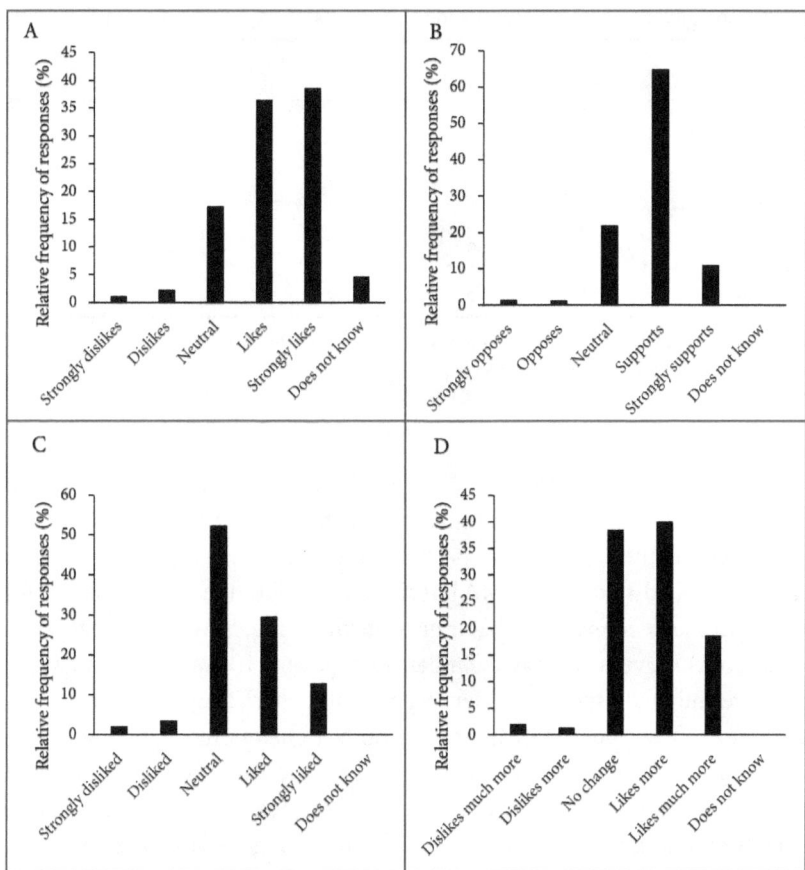

FIGURE 1. Community attitudes toward Randilen Wildlife Management Area based on questionnaire surveys administered in April–July 2020 to a randomly selected, proportionately weighted sample of 678 respondents from all eight member villages, using sub-villages as sampling strata.
A = General sentiment toward Randilen; **B** = Level of support for Randilen; **C** = General sentiment toward Randilen five years ago; **D** = Change in sentiment toward Randilen over past five years.

"normal life" in Maa (Woodhouse and McCabe 2018:5). While the relative place of neutral on the ordinal scale was made clear to respondents, this option can signify multiple things when given as a response. It can literally mean that the individual has neutral sentiments about the WMA, or it can mean that someone is choosing to withhold their honest or impulsive answers. This could be out of respect for leaders because of an aversion to disagreements and potential conflict or to avoid potential repercussions. In some instances, it

TABLE 1. Community attitudes toward Randilen WMA disaggregated by member village

	Strongly dislike	Dislike	Neutral	Like	Strongly Like	Does not know
Mswakini Chini	0	0	33	45	15	7
Mswakini Juu	2	0	35	57	5	1
Naitolia	0	0	24	41	33	3
Lemooti	5	7	5	2	76	5
Lengoolwa	3	2	11	10	72	3
Nafco	1	9	14	39	23	15
Lolkisale	0	0	0	35	63	2
Oldonyo	0	0	30	54	16	0

Numbers represent percentages of responses rounded to the nearest whole number based on questionnaire surveys administered in 2020 ($n = 678$).

is also used interchangeably with "I do not know" if the respondent wants to avoid the embarrassment of admitting that they do not know about something they feel they should. I suspect in reality there were many more people than zero who disliked the WMA but chose to withhold their true feelings for various reasons. Something that is also worth noting, however, is that Loveless's (2014) surveys generated similar homogeneity in responses, but at the opposite end of the spectrum. Her respondents generally opposed the WMA, and very few supported it, indicating both a change in people's attitudes toward the WMA and continuity in the tendency toward homogeneity. Based on my observations, it seems to me that rumors, fears, worries, and feelings of trust and appreciation spread like wildfire through Maasai communities, making it possible to track discursive shifts relative to preexisting baselines like Loveless's (2014) study.

The overall results reveal much consistency across the entire community, but there are a few telling details from each village's responses that are relevant to point out (see table 1). Arusha-dominated villages on the Naitolia side of the WMA were generally more "neutral" than Kisongo ones on the Lolkisale side, with the exception of Oldonyo. This finding suggests that there are still some lingering mixed feelings in the villages where Loveless (2014) documented discontent and protest. Though negative attitudes were not vocalized through the survey, the higher percentages of "neutral" responses relative to villages on the other side of the WMA are likely of anthropological significance. Overall levels of awareness about the WMA also seemed to be lower in those villages. Furthermore, Nafco had higher percentages of respondents with negative responses, and Lemooti had a majority who strongly liked the WMA, but a vocal minority who disliked it. Some ethnographic reasons for these differences are discussed in chapters 6 and 7.

Putting the Community First

The survey made clear my suspicions that support for the WMA was widespread across Randilen's member villages, albeit with some variations. One of the primary research questions that shaped my fieldwork in Randilen was, *Why* do community members support the WMA? I was particularly intrigued by the seemingly radical shift in attitudes that had taken place in the Arusha-inhabited villages of Naitolia and Mswakini. Were there specific social, political, and economic factors that had led to the emergence of positive conservation attitudes in a fashion that Agrawal (2005) would refer to as environmentality?

My ethnographic observations made clear that since Loveless's (2014) fieldwork, the WMA had developed into a livelihood-oriented conservation area that communities viewed as centrally important to their well-being. Interviews with community members revealed that people's fears about eviction and dispossession had given way to realizations that the WMA served the interests of local agropastoralists. As one Arusha woman (about thirty-nine years old) from Mswakini Juu articulated, "When the WMA was first established, we were afraid that we were going to be evicted from our homes. We were scared, and that made us dislike it. But since then, we have realized that the WMA is not like a park at all. We can stay in our homes and continue with our livelihood activities. I have seen how the WMA has helped us, and I like it now." Other interviewees reiterated this sentiment, including an Arusha man from Naitolia (about fifty-four years old) who explained, "There was a time that people hated the WMA because we were afraid that it might take all our land away. We came to understand that the WMA is ours because it is made up of the land of our village. We are allowed to graze our livestock in the WMA because it is our area. So, when we realized that the WMA was for our benefit, we started to like it." A key factor underlying the changing discourse across Randilen's member villages was that the title of "wildlife management area" was not mutually exclusive with "seasonal livestock grazing area." As one male Arusha elder (about ninety years old) from Mswakini Juu described during an interview, "We like the WMA because it is kind to our livestock. Those people who opposed it initially wanted that area for commercial farming. They wanted to keep their big farms in that place. As long as

Local cattle graze under the Randilen Wildlife Management Area highway sign near Naitolia village in 2024—a symbol of the way community members have come to view Randilen as a territory that supports local agropastoral livelihoods. Photo by author in 2024.

our livestock are grazing there, why would we dislike the WMA? The health of our livestock is what matters to us. So, we like it." As this man's explanation shows, people in Mswakini felt like WMA management cared about local pastoral livelihoods and the well-being of community members. Importantly, it was not pushing an ideology of pristine wilderness under the guise of conservation, nor had it implemented fences and fines that alienated the local community. Rather significantly, people felt like their concerns were meaningfully represented by WMA governance and management. Appreciation for Randilen was the norm rather than the exception.

Shifting conservation sentiments in Mswakini and Naitolia must be contextualized in relation to the institutional dynamics of the WMA and, in particular, the changing roles of nongovernmental organizations (NGOs) in Randilen's governance and management. In 2014, the African Wildlife Foundation's (AWF) contract with USAID ended, and they ceased work on the WMA. As they were closing up shop in Randilen, they reached out to Honeyguide, a grassroots Tanzanian NGO, with an invitation to facilitate anti-poaching activities in the WMA. Unlike other conservation NGOs, Honeyguide is a small and nimble organization comprising a team of around thirty Tanzanian citizens. Its main objective is to help communities participate in the wildlife sector through capacity-building for locally led models of conservation. This goal stems from an ethos that communities should be empowered to own and steward their land in ways that are in line with their customary ways of life. Honeyguide's community-oriented focus, which sets it apart from the big NGOs that often dominate the conservation space in Tanzania, was born of the organizational founder's experiences establishing one of the first photographic village-based tourism ventures in Ololosokwan, Loliondo ("Sokwe Camps") that made clear the possibilities of mutually enhancing conservation initiatives that benefit local communities through devolution of authority. Honeyguide's five main programmatic aims are helping communities ensure the economic viability of local conservation areas, building capacity for governance and management, facilitating communication between communities and conservation authorities, reducing human-wildlife conflict, and safeguarding rangelands and habitat for community livelihoods and wildlife.

Prior to working in Randilen, Honeyguide had developed a capable anti-poaching track record in Enduimet WMA and was contracted by AWF to carry out conservation enforcement on Manyara Ranch, a semi-community-based conservation area adjacent to Naitolia and Mswakini on the northwestern side of the A104 Arusha-Babati highway (Goldman 2020). At the time (2012–2013), Manyara Ranch was having a difficult time curbing elephant poaching. Honeyguide successfully stopped the surge in ivory plundering on the ranch in part due to its canine anti-poaching unit, which helped track perpetrators all the way back to their homes. AWF had suggested that Honeyguide extend its operations into Randilen WMA to create a larger presence in the area and streamline management activities across the ranch and the WMA. Unlike its agreement with AWF on Manyara Ranch, however, Honeyguide signed its contract in Randilen with the community-based organization (CBO) and subsequently applied for donor funds from World Wildlife Fund (WWF) and USAID directly to carry out its anti-poaching operations in the WMA. After learning through experience in Manyara Ranch and Enduimet WMA the importance of strong local governance institutions, Honeyguide worked with Randilen's CBO and encouraged them to establish a functional, robust, and democratic Authorized Association (AA) that genuinely represented their community's interests. Thus, rather than the typical model of an NGO projecting its vision of conservation onto a local setting, Honeyguide's starting point was ensuring that the community was well positioned to make its own decisions and steer the WMA in a direction of its choosing. This meant focusing on the AA and ensuring that it had the tools to function effectively. After Honeyguide helped the community establish its institutional framework for governance—a representative AA and steering committees—the community decided to work with Honeyguide to build up management capacity of the WMA. Honeyguide's role was to help the community operationalize the WMA in a way that fulfilled local goals and aspirations, and this meant that Honeyguide remained directly accountable to the community rather than outside interests. This institutional structure gave Honeyguide and the AA a significant degree of liberty to devise a management plan on their own terms.

With the departure of AWF and the arrival of Honeyguide, the sociopolitical dynamics of the WMA began to change markedly for the better. As I have since explicated through my ethnographic fieldwork across Randilen's member villages, the WMA's promising developments have largely been a product of Honeyguide's concerted efforts to put community livelihoods on equal footing with wildlife conservation. To contextualize Honeyguide's dedication to local interests, it is important to track its historical evolution as an organization. Damian Bell founded Honeyguide in 2007 to build bridges between tourism and communities that were mutually beneficial for people and wildlife. While "win-win" models of conservation in Tanzania had been elusive, Damian's past experiences as a tour operator in Ololosokwan in the 1990s instilled in him an appreciation for what was possible. His initial company, Sokwe, had developed a well-respected reputation for negotiating with local Maasai in good faith to establish photographic concessions on village land in exchange for fair returns and guaranteed livestock grazing rights. Damian was thus aware that tourism had the potential to benefit communities if it was organized in a responsible way. In Honeyguide's early days, its projects were diverse and innovative. Honeyguide experimented with a food production project in the western Serengeti to help bushmeat-dependent communities sell their produce to tourists through a community-based market and business center. In the eastern Serengeti, it helped create tourism management plans for adjacent Maasai communities. Near Longido, Honeyguide collaborated with the Big Life Foundation, a Kenyan conservation NGO focused on reducing elephant poaching in the Amboseli-Tsavo-Kilimanjaro ecosystem with funding support from AWF.

While international donors coveted anti-poaching initiatives as the silver bullet of wildlife conservation projects in East Africa, Honeyguide's interests were broader and more tailored toward enhancing the sustainability of community-based conservation. Big Life Foundation, like many NGOs in East Africa, was keen on branding itself as the leading anti-poaching organization in Amboseli and employed a militarized and heavy-handed approach to conservation enforcement. They had secured significant funding in support of their activities and believed unwaveringly in "pumping money" into anti-poaching efforts in Enduimet WMA. Honeyguide's views on anti-poaching, however, were beginning to diverge from big conservation NGOs. As Damian described in an interview in 2020:

> We didn't have much poaching [in Enduimet] after a short while; it was more human-wildlife conflict. The incidents were mostly around lions where some Maasai would go and kill a lion because it killed a cow or something. And [Big Life] would come in and say, right, arrest those people; arrest anyone! They literally came in and wanted me to just go arrest people and throw them in jail. Right, that'll stop them! *laughs* Like, that's not my job. But they were really serious about it.... So there was a bit of [tension]; they weren't happy with our approach. They were also not happy with the fact that when there were problems with the community rangers, that we didn't come in and just fire the ranger. Well, I'm not employing the ranger, the community is! And there were issues when they wanted us to move rangers from one place to another because they wanted more control, but I was trying to build the capacity of the communities. So there was a difference in management styles.

As Damian's narrative segment reveals, Honeyguide was wary of anti-poaching efforts that hinged on policing the community in a top-down fashion. Honeyguide was interested in forging a new direction that built up the capacities of communities to manage wildlife themselves. These values are exemplified in the organization's name, which takes after the Honeyguide savanna bird. The Honeyguide bird is known for leading Indigenous hunters to beehives where they can find honey, providing the birds with wax and larvae to enjoy in exchange after the bees are smoked out. The symbiotic relationship between people and honeybirds is a metaphor for Honeyguide's collaborative rapport with communities. While Honeyguide and Big Life found themselves moving in different directions, Honeyguide's experiences working in Enduimet proved to be of great value in helping the organization refine its vision of community-based conservation through a practical set of experiences. One of the key lessons that Honeyguide learned was that traditional anti-poaching efforts not only instilled resentment in local communities, but they were also expensive. These issues became clear when Damian started to dig into the economics of ranger posts. As he explained:

> Once a ranger post is constructed, you need a minimum of three people to guard it, because you've got rotations. It's three because you always need two people at a time in a ranger post. You're not allowed just one, so it's two on

and one on rotation. So you have to have three people guard a ranger post, then you've got it sorted. Now you have to make it useful. So then you have to have another seven people based at the ranger posts to do activities around and go on patrols—walking around or driving around or whatever they do. So now you've got ten people per ranger post, and Enduimet had five ranger posts. Well, that's fifty people in just the ranger posts! Then you've got your drivers on top of that. So suddenly you've got up to sixty people on the anti-poaching team. And suddenly you're talking about money [and realizing this adds up fast]. So that's when I started thinking, okay, we've got to rethink this model. There is absolutely no way we can carry on. And that's when I started thinking, well, what's actually successful out there? Why are we doing all this ranger posting stuff? And that's when I thought, we need to rethink the whole model. This is ridiculous. We're building an anti-poaching unit not on the threat of poaching, but on a concept whereby you have to have put rangers in a space just because that space has wildlife. We need to protect this area, so bang a ranger post in there. That was the attitude! Rather than say, is there a threat of poaching? Do we need to throw money in there?

Honeyguide's experiences overseeing the anti-poaching units in Enduimet led the organization to develop some degree of cynicism about the practicality of the classic anti-poaching model of conservation that was based on monitoring a vast area through ranger patrols. The anti-poaching model of conservation had taken its inspiration from national parks but was missing a key factor in the context of WMAs: community interest and buy-in. In Damian's description:

If you think about how patrols got set up, I mean, what's the basis of patrols? I guess it comes from national parks with no people in them where you've got rangers who need to go on patrol because there's nobody there. They're looking out for footprints, and if they find footprints, it means a person's been there on foot, thus it must have been a poacher because in national parks you're not allowed on foot inside them. And then obviously when conservation went outside national parks, they picked up that same concept and threw it on them. This is the way: you wear green, you put black boots on, you salute, you carry a gun. Okay, and you go around and arrest people who shouldn't be there. *Laughs* And you suddenly tell communities to do exactly the same thing. That's the way it's done! And obviously, some of them kind of like to do that. It's kind of fun. You know, everyone's got their thing. You get a sense of pride. But if you bury that and say, well, what am I trying to achieve? Well, fundamentally, there are other interested parties like national parks, NGOs, and investors who want community land to remain as natural habitat for wildlife. So you are asking communities to stop converting it primarily. That's the biggest threat. Stopping poaching is what everyone thinks you have to do. But actually, the main issue is stopping habitat from being converted into farmland. So you want them to protect that habitat and the wildlife within it. Well, if you just focus on that, then you've got to rethink things. What do you need to do in order to get there? Well, the biggest thing is to make sure that everybody in that community wants [the WMA] to be there! Focus on that!

As time passed in Enduimet, Damian thought deeply about the issues affecting community-based conservation, and Honeyguide's focus began to shift away from classic anti-poaching efforts toward things that were important from the perspectives of communities. It was becoming clear that bringing an external body into communities to police them was far less productive than helping communities realize the value of wildlife on their own terms. As Damian thought reflexively about the merits of traditional anti-poaching efforts, Honeyguide began to come into its own and evolve as an organization. In 2013, it commissioned a series of surveys in Enduimet to understand what communities cared most about in the context of conservation. They found that communities considered their agropastoral livelihoods to be of the greatest importance, particularly in the context of human-wildlife conflicts. Given the wealth of wildlife in the Amboseli ecosystem, people were concerned mainly about crop-raiding elephants and livestock depredation from lions and other carnivores. Interestingly, Honeyguide found in their survey that the money coming in from tourism did not feature as high on the community's list of priorities as expected. The main concern from the perspective of local communities was reducing the livelihood costs of increasing conflict with wildlife. This suggested to Honeyguide that communities might be willing to reinvest a significant portion of their tourism revenues into the operational costs of a WMA if it assisted in reducing the burdens of living with wildlife.

After carrying out the community surveys in Enduimet in 2013, Honeyguide commenced work on human-wildlife conflict reduction in Burunge WMA, which the participating villages supported. Honeyguide rallied a small team of thirty volunteers to work together on people's smallholder farms to help defend them from crop-raiding wildlife. Unfortunately, Burunge WMA had already become accustomed to large international NGOs coming in, carrying out projects, and leaving, with no real plans for how projects would be continued after the NGOs left. Honeyguide, by contrast, was interested in building an initiative that communities could manage on their own into the future through the WMA's conservation model. Burunge WMA's leadership team, however, immediately discontinued the project after Honeyguide left because they were "more interested in making money" than in delivering services to the communities that local people actually valued. Although Burunge WMA did not continue the project after Honeyguide left, the initiative showed great promise in terms of community attitudes because the participating villagers viewed it as directly valuable to their livelihoods. The community buy-in was there, but the sustainability was not, and some of these lessons were once again internalized by Honeyguide as it transitioned to work in Randilen in 2014.

When Honeyguide arrived in Randilen, the WMA was essentially a "paper reserve" that existed in law but had no tangible presence on the ground. While AWF had assisted in hurriedly setting up the initial boundaries of the WMA and in preparing the initial zoning plans, there was no main office, no infrastructure, and no entrance gate. This was a blessing for Honeyguide, as it meant that Randilen was a blank slate to work with. Unlike Burunge and Enduimet, where numerous international NGOs had existing stakes, Honeyguide could start fresh in Randilen in collaboration with the community. Though AWF, Big Life, Ujamaa Community Resource Team, and the Nature Conservancy had all partnered with Honeyguide in various ways, at that point, Honeyguide was the only organization actually working on the ground in Randilen WMA. Thus, while AWF had been involved in Randilen's setup on paper, management in practice had yet to unfold. Consequently, the community had not been indoctrinated by what Damian facetiously described as "overgenerous NGOs," which were inclined to carry out top-down projects to fit their deliverables without thinking critically about the community's long-term interests. In Enduimet, for instance, WWF had funded the construction of four entrance gates but had not taken into account the long-term running costs of securing them from looting, let alone the money it would take to systematically manage their operations. Damian estimates it would take twenty people to manage those gates alone, which by way of comparison, is just shy of Makame WMA's entire anti-poaching unit (twenty-seven people at the time of writing). Considering this, when Honeyguide entered Randilen, it was very cautious to ensure that it did not invest anything that would burden the community with long-term costs when it left. This dovetailed with Honeyguide's interests in leaving the community in a more empowered state than when it first began. In 2016, Honeyguide refined this ethos in its organizational strategy, which articulated the need for carefully planned exit strategies to promote long-term sustainability. Rather than parachuting in and formalizing WMAs across a vast area as quickly as possible, as AWF had, this type of approach entailed thinking thoughtfully about how Honeyguide could help shepherd a community-based area to a point where it was financially sustainable and appreciated by its local community. The key to reaching this goal was building something that could stand on its own for the benefit of its community.

New Directions for WMA Leadership and Administration

When Honeyguide commenced work in Randilen in 2014, there was no real political leadership of the WMA. Though an AA existed on paper, it had not yet had a proper governance meeting to decide Randilen's management priorities and future directions. The WMA had no ranger posts or employees, and management comprised only a few volunteer village game scouts (VGS) from the member villages. Because there were no big NGOs that had been involved in influencing Randilen's governance and management dynamics, as had occurred in Burunge and other WMAs, Randilen's CBO was very receptive to Honeyguide's proposal to help craft the WMA into a sustainable, community-based enterprise through institutional reform.

Honeyguide's aim was for the community to take hold of all aspects of the WMA's governance and management, including the handling of finances. This meant revising the institutional structures of the WMA. Drawing from Damian's past experience as a tour operator, Honeyguide focused on helping the community realize the financial viability of the WMA by running it more like a business. This is a key aspect of Honeyguide's approach in Randilen that is often lacking in the context of WMAs—detailed attention to how the business can be run in an efficient and sustainable way. Honeyguide also helped establish clear distinctions between the WMA's governance board and management team after observing in Burunge WMA that messy overlaps between the two can undermine efficiency and accountability. Separating them in Randilen has meant clearer roles for WMA stakeholders, allowing everyone to carry out their duties more effectively. Randilen hired Meshurie Melembuki, who became the first professional manager of a WMA in Tanzania, and he has done an excellent job of spearheading Randilen's community-based management initiatives. Meshurie understands ecological dynamics through his education at the College of African Wildlife Management Mweka, but more importantly, he is an Arusha agropastoralist from Mswakini Juu who is straightforward in his dealings, and well respected by the local community. Meshurie has a sound grasp of the economic challenges of administering a conservation area, which makes him a practical and responsible manager. But perhaps most impactfully, he is a critical thinker who recognizes social-ecological complexity and the need to revisit conventional thinking about the relationships between conservation and pastoralism. As he described to me during a conversation in mid-2022:

> These conservationists have a dead way of thinking. The national park services that TANAPA provides to communities should not be money but livestock grazing access. Do not talk about environmental degradation. Talk about management. Sixty percent of Tarangire's wildlife is outside the park. Why would the wildlife seek out those areas if they have been overgrazed and overutilized by the local communities? If 60% of Tarangire's wildlife is outside the park, then the conservationist theories of how nature should look are wrong. The problem is that we are still stuck in the mindset of colonial conservation. It is a threat to Africa today. We need a holistic conservation model that balances all that we have in a mutual way.

Meshurie's call to abandon the fortress model of conservation is reflective of his personal management philosophy that conservation should work in a holistic way *together* with local communities. This ethos shines through Randilen's management practices, which approach local pastoral livelihoods as an integrated part of the WMA's conservation model.

To further round out the administrative team, Randilen hired Samwel Saruni Mollel, a professional accountant from Lengoolwa, to manage the books. Employing a professionally trained finance and administration team comprising highly qualified personnel from the local community has been a groundbreaking new direction for WMAs in Tanzania that greatly enhances local capacity for community-based conservation. Honeyguide is keenly aware of the fact that conservation areas are complex businesses, and even large-scale organizations like African Parks struggle to manage them efficiently. And yet, local communities are often expected to navigate the challenges of operating multidimensional businesses in collaboration with investors, governments, and NGOs without adequate support and training. Most WMAs have limited or nonexistent financial, human resource, and administrative policies, making it difficult for people who have not been trained in those fields to develop skills on the fly. Realizing this, Honeyguide focused its efforts on capacity-building to help the Randilen community cultivate technical expertise through coaching, education, workshops, and trainings.

Building Local Capacity for Governance and Management

To ensure that Randilen's management priorities aligned with the interests of the community it represented, Honeyguide also keyed in on the importance of WMA governance. One of Honeyguide's core strategies was empowering Randilen's governance body to function independently of NGO assistance through capacity-building initiatives. The rise of local support for Randilen since 2015 has largely been tied to the emergence of equitable and devolved WMA institutions for taking decisions regarding the management of community land and wildlife.

In many ways, the WMA's core governance institutions mirror those of village councils but operate at a higher political scale since they encompass multiple villages. However, WMA institutions do not exist in a political vacuum and are layered in relation to existing village-level ones. Village institutions continue to play a significant role in determining *who* will represent each village at the WMA governance table, and *how* WMA revenues will be used in each member village. Thus, village governments are gatekeepers who determine degrees of community participation in WMAs and the flows of benefits from them.

The main governance body of the WMA is the AA, which in the case of Randilen comprises five people from each of its eight member villages. The WMA also has two independent groups that provide guidance and oversight to the AA but are not directly responsible for taking governance decisions: The District Advisory Board and the Board of Trustees. The District Advisory Board assists the AA in coordinating and collaborating with external stakeholders like government and NGOs but does not provide direct governance input relating to WMA management. It was established in accordance with section 33(1) of the Wildlife Conservation Regulations of 2012 and is made up of appointed district government officials. The Board of Trustees comprises one individual elected from each of Randilen's member villages and is tasked with providing independent governance oversight to hold the AA accountable in instances of potential conflicts of interest.

All WMA governance decisions are taken by the AA. While Loveless (2014) pointed out that the initial AA members in Naitolia and Mswakini were cherry-picked by village committees, Honeyguide helped ensure that the AA grew into a more robust and effective institution that genuinely represented the interests of its community. After commencing work in Randilen, Honeyguide facilitated a three-day workshop that brought together all AA members and provided education on the nuances of WMA legislation, the roles of AA members, and the particularities of WMA governance and management. The main objective of these trainings was to enable the WMA to work for its community through a strong and devolved governance institution. Village council members were also educated about the importance of engaging meaningfully with the WMA through the AA institution to ensure that WMA governance decisions reflected community interests.

Building from these initial workshops, Honeyguide facilitated a series of trainings that enhanced communication between WMA governance actors. Most recently, Honeyguide organized a "site-level assessment of governance and equity" workshop in November 2019, which I contributed to as focus group discussion facilitator. With support from the International Institute for Environment and Development, the event brought together AA members, village and district government officials, representatives from tourism camps, NGOs, and Randilen's management team. These sets of stakeholders were divided at the workshop and each subsequently scored the WMA on a series of governance-related measures, including the rights of community members, gender dynamics, discrimination, participation, transparency, accountability, and social justice. All workshop attendees then came together to discuss measures where the groups reported a dissonance in their views. My sense of this governance workshop was not that it was a cookie-cutter exercise of trivial value (as is sometimes the case with donor or grant-funded workshops), but one that was very generative for the meeting attendees who animated the workshop with their understandings of Randilen's local sociopolitical context.

Through Honeyguide's efforts to improve Randilen's governance capacity, the AA has developed into a strong institution that allows the community to shape the WMA's management direction. Randilen's AA members serve five-year terms and have the formal authority to govern the WMA on behalf of the community pursuant to the Wildlife Conservation Act No. 5 of 2009 and the most recent WMA Regulations. AA candidates are selected democratically at the village level through village assembly meetings. Candidates are chosen for a variety of reasons, ranging from their popularity within villages, their dispositions, their education level, and perhaps most significantly, their skills as communicators. From a community perspective, AA representatives are the key go-between for villages and the WMA since they attend Randilen's governance meetings and provide updates to their village assemblies upon their return to their respective villages. As described by one male Arusha AA member (about thirty-five years old) from Lengoolwa, "I am happy to be an AA

Randilen Wildlife Management Area authorized association members participate in a subcommittee governance meeting in Makuyuni village. Photo by author in 2020.

member because I like communicating with the representatives from other villages and knowing what benefits we are getting from the WMA. I was chosen by my people to represent them. I like attending the meetings and listening to whatever the WMA is doing and then bringing back the knowledge I learn to my village. Our main task as an AA is making sure that the WMA represents the interests of each member village equally, not one village over other villages, but our community as a whole." Once AA members have been selected from each village, they hold a vote to determine who will serve as the WMA chair and vice chair. From 2017 to 2022, the chair was an Arusha man from Mswakini Chini, and the vice chair was an Arusha woman from Oldonyo. For the period of 2022–2027, the WMA chair was an Arusha man from Lengoolwa, and the vice chair was from Naitolia. AA members are grouped into subcommittees, which are tasked with overseeing particular segments of WMA governance.

At the time of my doctoral fieldwork in 2019–2020, the AA had three subcommittees: a protection committee that focused on the design and enforcement of environmental regulations, a finance and planning committee that oversaw the budget and strategic plan for the year, and an executive committee that received the resolutions of all other committees and assessed their decisions independently. Eight AA members sat on each subcommittee, with one individual representing each village. This meant that twenty-four members participated in small committee meetings, but sixteen did not. In 2023, the AA added two new subcommittees—a community committee that discusses issues relating to community development like the distributions of scholarships, and a rangeland committee that addresses the management of WMA pastures and decisions around seasonal grazing access. The addition of two new subcommittees was agreed upon by the AA as a way to further enhance community participation by ensuring that all AA members actively participate in WMA governance. The new group structure is tailored to the overall size of the AA with eight members sitting on all five subcommittees. Each of the WMA's eight member villages thus has one member on every committee, and every AA member is part of a committee. To assign AA members to committees, each group of five village representatives sits together and decides on two preferred subcommittees for each person. The wider AA then gathers and votes on which of the two subcommittees the individuals are assigned to, usually considering gender balances of each committee. This decentralized political system enables AA members to participate in WMA governance in a meaningful way while also ensuring that each member village is fairly represented.

Randilen's AA subcommittees synergize with existing village-level institutions to ensure that local herders are closely involved in shaping land use plans and rangeland management practices. Grazing committees in the member villages serve as a direct liaison between villages and WMA management staff by holding regular meetings with the rangeland committee of the WMA to voice their concerns about current herding conditions. These meetings also provide an opportunity for grazing committees to learn about current developments in WMA management emerging from recent AA meetings. The bidirectional communication between local grazing committees and the WMA's rangeland committee ensures that herders are aware of the WMA's conservation regulations, are involved in shaping the enforcement of livestock grazing plans, and are represented by the WMA's governance and management dynamics. Close interactions between the WMA and local grazing committees allow herders to feel heard by conservation authorities and help keep pastoralist livelihood concerns at the forefront of conservation planning. As described by the chair of the local livestock grazing committee in Mswakini Juu during an interview:

> As the local chair of pastoralists, I am the one to organize meetings between the WMA representatives and our herding committee. We have nine mem-

bers of the committee who were selected during the village assembly meeting to represent the village as the whole. Those members nominated me as the chair of our committee and together we serve for five years. During our meetings with AA members and WMA staff, we explain our issues to them and they inform us as well about the conservation regulations. We agree together that these are the months that we will not be allowed to take our livestock there so that we can let that area recover and conserve that grass in case of drought. So we get a lot of benefits from the WMA because our livestock are grazing there. The WMA is within us.

As this narrative segment illustrates, village-level institutions for rangeland management are brought into direct conversation with the WMA's rangeland subcommittee to ensure that local herders are involved in WMA governance and have a stake in the management of the area. Conservation enforcement in Randilen is thus not at the expense of its community through a "fines and fences" approach, but is implemented in a way that represents the core Maasai interest in managing common pastures for community well-being.

General AA meetings are held on a quarterly basis, usually in Makuyuni or Mto Wa Mbu, and are attended by Randilen's management team, AA members, the Board of Trustees, and members of the District Advisory Board. I attended several of these meetings in 2019 and 2020 and noted devolved and inclusive discourse on each occasion. Notably, Honeyguide representatives were not present at those meetings—governance was in the hands of the community. All AA meetings begin with a chant led by the WMA chair:

Tembo, na maendeleo; Maendeleo, na tembo
Mifugo, na maisha; Maisha, na mifugo

(Elephants, and development; Development, and elephants
Livestock, and life; Life, and livestock)

The verse's unifying theme encapsulates the ethos of Randilen WMA as a community-based conservation area that manages wildlife and promotes safari tourism while simultaneously sustaining its local pastoral community by providing pasture for livestock. Elephants are linked to development through the distribution of tourism revenue and livestock is likened to life itself.

Following the opening chant, AA meetings begin with a gathering of the whole to brief attendees about current developments, key organizational matters, financial reports, and emergent issues. During this session, all AA members are invited to ask questions and introduce topics that they would like to discuss with the group. This provides an opportunity for people to present concerns on behalf of their villages, since AA members are vessels for communicating community perspectives to the WMA's governance body. The meeting of the whole is then broken down into subcommittees to discuss specific issues related to WMA governance. Subcommittees sit in circular formations and participate in decentralized breakout groups. They then return to the larger meeting to discuss their findings and reflect on decisions as a collective. Depending on the number of issues to be discussed during subcommittee meetings, they are sometimes scheduled on different days in advance of the main meetings.

I attended a quarterly AA meeting at a conference hall in Mto wa Mbu on April 30, 2024, during which the management team presented Randilen's operating budget and financial reports and discussed the WMA workplan for the rest of the year. I took field notes of the event and provide some thick description here to contextualize Randilen's institutional dynamics. Fifty-five people were in attendance, including AA members, representatives of the advisory boards, management staff, and children accompanying female AA members. During the opening parts of the meeting, all AA members sat facing the WMA chair, vice chair, and manager seated at the front of the room. AA members took turns raising issues of significance to their everyday lives. In this case, heavy rains between November 2023 and April 2024 had led to increased dispersals of elephants on village land, and community members expressed concerns about the need for greater support from the WMA to protect their farms. The chair and manager responded to each point thoughtfully with suggestions about how the expressed concerns could be addressed. In this case, it was suggested that the WMA would increase human-elephant conflict patrols as the budget allowed. Each point was discussed openly as a group until a satisfactory resolution had been reached. The meeting jumped between kiSwahili and Maa, and both women and men stood up and shared their points of view. Young children sat with their mothers, allowing women to participate equitably in the meeting—a key difference from the exclusionary gender dynamics documented a decade earlier by Loveless (2014). I observed that the group was focused on the task at hand. Individuals were not filtering in and out to receive phone calls or attend to other matters. Randilen's AA members were actively engaged participants.

The meeting then moved on to its primary objective, and financial statements and operating budgets were projected on a TV screen at the front of the room. Any AA member had the opportunity to stand up and pose questions, raise issues to be discussed, and express their opinion on the WMA's financial statements. A key feature of Randilen's governance structure that sets it apart from other conservation areas is that the AA holds the authority to determine the WMA's budget and management workplan. This means that the AA is intimately aware of Randilen's financial statements and revenue flows and has the ultimate say in determining how the WMA uses its funds. AA members are made aware of the basic WMA running costs and can decide to shift allocations as they see fit. In this case, the participants were content with the current use of WMA funds and the WMA's strategic vision for the rest of the year. Given the time-intensive nature of this meeting, subcommittee meetings were scheduled for different days.

The cumulative effect of Honeyguide's efforts to help build Randilen's capacity for governance has been an AA that equitably represents its member villages and collaborates efficiently with NGOs, investors, and external government actors at different scales. Devolved AA meetings ensure that the land is managed in a way that benefits the community as a whole and is not subjected to fragmentation or private interests. Unlike the village institution, which is at risk of corruption and nefarious dealings (chapter 7), governance authority of the WMA is distributed across the eight villages represented by the AA and is thus difficult to co-opt by individuals. Local appreciation for the WMA's governance dynamics is ethnographically visible through participant observation and qualitative accounts from community members. Further triangulating these findings, my quantitative research showed that 92 percent ($n=624$) of survey respondents across Randilen's member villages felt that their community was included in WMA governance, a shining indicator of conservation equity. Good governance seems to be a very crucial reason for Randilen's success in garnering local support for conservation.

Once governance decisions have been taken, Randilen's management team enforces them in practice. The WMA's head manager, Meshurie, follows the directives established by the AA to enact a management workplan. Meshurie oversees four distinct management units within the WMA: a protection unit, a tourism unit, a community unit, and a financial team. The community unit is headed by the WMA chair and is responsible for addressing issues that the community feels to be of particular concern as conveyed by the AA. The tourism unit is responsible for handling visitor permits, communication with investors, and the operation of the entrance gate. The financial unit works on the WMA's operating budget under Samwel's professional accounting leadership.

Protection activities, including on-the-ground enforcement of WMA regulations, are carried out by VGS who are hired exclusively from an applicant pool of community members. In total, Randilen has twenty-six VGS, with three to four selected from each member village. VGS are then disaggregated into a camp group, a zonal group, and an entrance gate group, with zonal VGS considered part of the protection unit and ones staffed at the entrance gate part of the tourism unit. Those stationed at the main VGS camps in the WMA are on-call in the event of issues as they arise. By securing international donor funds for anti-poaching operations, Honeyguide trains, funds, and provides guidance for Randilen's VGS but is not directly responsible for managing the WMA. Conservation enforcement is carried out by VGS in collaboration with Randilen's management team and in a way that represents the interests of the AA. The fact that VGS are from member villages plays a major role in fostering community trust in WMA management and in ensuring that WMA management is implemented in a way that is empathetic to community concerns. VGS are themselves Maasai herders and farmers from the member villages and thus share people's livelihood concerns. This allows them to exercise discretion and flexibility when enforcing bylaws in ways that are adapted to the local ethnographic context. VGS strictly enforce transgressions on WMA regulations from elite herders from outside the local community, as well as instances of poaching, but they are sympathetic to local livestock grazing concerns, especially during times of stress. This form of adaptive management and flexible enforcement of boundaries ensures that the WMA serves the interests of local livelihoods while adequately protecting wildlife and preventing outside actors from grabbing or encroaching on community land. Significantly, my quantitative results revealed that 91 percent ($n = 620$) of survey respondents across Randilen's member villages felt that WMA management practices equitably represented community concerns.

The benefits that formalization offers local pastoralists cannot be understated. Codification of the WMA as a formal institution supported by state wildlife laws ensures that the tenure status of Randilen's communal grazing area is secure. Strategically reclassifying Lolkisale's livestock grazing area as a WMA means that it is unlikely to be grabbed by the state and reclassified as a centrally managed protected area, despite its significance for wildlife. Although TAWA first collects revenue from the WMA before subsequently distributing it back, it is not involved in the governance and day-to-day management of Randilen WMA. Thus, many of the fears expressed by the Maasai of Loliondo (chapter 1) around the prospects of a WMA being used by the central government to displace pastoral communities have not come to be in Lolkisale. Randilen WMA has in fact done the opposite: it has formalized a locally led institution for managing land and wildlife that is recognized and respected by state law. Unlike Certificates of Customary Rights of Occupancy

(CCROs), which provide tenure security for villages but do not afford legal rights to manage wildlife, Randilen WMA provides its community with state-sanctioned wildlife user rights and formal authority to manage a multiple-use conservation area in a key wildlife dispersal area. The risk of Randilen's member villages being evicted in favor of a game reserve is now low, even despite the ecological significance of the area for Tarangire's elephants. Being able to enforce their local zoning plans with patrol vehicles and armed VGS allows the community a layer of protection for their common pasture that they would not have been able to secure through CCROs alone. Within Randilen WMA, Weber's "monopoly on the direct use of force" is devolved to VGS, empowering them to defend pastoral territory from elite outsiders seeking to appropriate common pastures for personal gains. Randilen WMA thus equips herders with the tools to protect their pastoral way of life.

CHAPTER SIX

Foundations of a Social Enterprise

The key to unlocking community support for Randilen has been the orientation of the conservation model toward things that local people care about. Understanding what community members value, and why, has demanded a dedicated effort to take the point of view of the people who are directly affected by conservation practice. I conceptualize this process as empathetic conservation.

Honeyguide helped foster a management strategy that directly improved the everyday lives of its community members by safeguarding local agropastoral livelihoods. Wild animals, while rewarding to observe and photograph in the context of safari tourism, become far less attractive when they prevent children from attending school, attack livestock, raid crops, and kill loved ones. Expecting people living with wildlife to value them in the same way as safari-goers is unreasonable given these costs. Randilen's leadership team prioritized distributing the benefits of the Wildlife Management Area (WMA) to individual households by considering what was important in the context of people's livelihoods. Honeyguide keyed in on two main economic practices that community members were concerned about: farming and livestock grazing. While the Arusha care more about their farms than the Kisongo, they also value their livestock. Conversely, though the Kisongo still consider themselves "people of cattle," they have diversified their economies to include farming in the face of constraints on the pastoral mode of production. Honeyguide was able to establish, in what I would call an anthropological fashion, that what mattered most to livestock keepers was *grass*. As described by a Kisongo man (about forty-six years old) from Oldonyo in 2020, "We pastoralists love the WMA because when it first came, there were big farms that were there. The WMA took them and converted them to communal grazing land. There were some rich farmers who were shouting about losing their farms and

An Arusha woman in Oldonyo village threshes beans with a stick. Photo by author in 2020.

trying to advise people to oppose the WMA. But we do not want that place to be a private farm. We need it to be a grazing area for the benefit of the whole community." As this herder's narrative shows, local livestock keepers have come to see Randilen as a means of securing pasture at the expense of wealthy farmers whose large private holdings threatened to enclose the pastoral commons. Other Kisongo pastoralists consistently voiced their appreciation for the way that Randilen conserves grass for local livestock. In the words of a Kisongo man (about forty-eight years old) from Lengoolwa, "I love Randilen WMA very much. Why? Because it is ours. The grass is kept for our livestock, and we get some money from the tourist activities to support our community." Herder affection for Randilen is rooted in an understanding of how the WMA fits in relation to pastoralism. A resonating theme across interviewees was that herders see the WMA as an extension of their pastoral territory that is managed in support of the livestock economy.

In terms of cultivation, smallholder Arusha farmers wanted secure tenure rights and assistance in keeping elephants and other wild animals out of their farms. Through concerted efforts, Honeyguide was able to reduce crop damage in member villages considerably. This involved recruiting about three hundred volunteers from member villages to assist in monitoring the expansive farm areas in Nafco, Naitolia, and Mswakini. While Honeyguide could not offer salaries for all these people, the participants were happy to be involved and take ownership of an issue that directly affected their livelihoods. In essence, these efforts reflected a form of management decentralization that allowed people to feel included in the enforcement of the WMA. Participatory management institutions have been shown in political ecology literature to directly affect local attitudes toward conservation areas (Agrawal 2005).

To further bolster these efforts, Honeyguide subsidized human-elephant conflict (HEC) toolkits comprising defensive equipment to help farmers protect their crops. These included flashlights, air horns, roman candles, and chili bombs, which were not meant to be used as weapons against the elephants, but as deterrents to prevent the elephants from dwelling in people's farms. Randilen's community members deeply appreciate these efforts to reduce the impacts of elephants on crop production. In the words of an Arusha woman (about forty-eight years old) from Mswakini Juu, "The WMA staff come to help and give us instruments to defend our farms from wildlife like torches and chili bombs, so we are very grateful to them. They are good neighbors that help us defend our farms and build schools, so we take care of Randilen because it takes care of us." As this woman's explanation reveals, the provision of HEC toolkits fostered a sense of reciprocity with local Arusha, who took the initiative to mean that WMA management cared about the livelihoods of smallholder cultivators. As her narrative makes clear, conservation dynamics are about a *relationship* between managers and community members. Feelings of being cared for often organically give rise to reciprocal feelings of respect and appreciation for the caregiver. Fundamental to this is the recognition and validation of community livelihood concerns as things worth caring about. I posit here that empathetic conservation is less about charity than it is about humility, reflexivity, and open-mindedness, manifest as effort to learn about people, their lived experiences, and their cultural values. Having *conservation in common* is not about molding people into hyperrational conservationists through governmentalizing discourse and institutional reform, nor is it about excluding people entirely by forming fortresses of enclosed "wilderness." Rather, it is about taking initiative to care about the things that community members care about and contextualizing the concept of conservation in a locally meaningful way.

Equipping local cultivators with HEC toolkits has significantly helped to reduce the costs of wildlife on local livelihoods (Raycraft 2023, 2024a). As I witnessed during fieldwork, when farmers did not have the equipment on hand, they usually yelled and banged on buckets until the elephants left. While the elephant sympathizer in me wondered whether aggravating the animals in this way was the most effective way to promote coexistence, as an anthropologist, I could not help but empathize with the plight of smallholder maize farmers in the Randilen WMA member villages who were otherwise ill-equipped to safeguard their crops from herds of the largest land animal on Earth. Over the course of my fieldwork, dozens of people were injured by elephants, some severely so. A few were even killed. Many of the Kisongo I interviewed in Lemooti had abandoned planting their farms knowing that they were fighting a losing battle with elephants. Nafco, Naitolia, and Mswakini Juu were also heavily impacted by elephants, but the Arusha in these villages were determined to remain farmers, which they viewed as central to their economy and way of life. Rather than devising a business model for Randilen that was purely about generating monetary returns for communities, Honeyguide realized that the WMA should be catered toward addressing local livelihood and well-being concerns and generating enough money to sustain itself in the process. Money to build schools and dispensaries is very much appreciated by community members but constitutes the icing on the cake compared to people's primary livelihood concerns.

Rethinking the business model was an exercise in empathy, and one that allowed Honeyguide to address the things that community members felt mattered to them. It meant revising the business plan in terms of what the community was investing in the WMA and what it was getting in return. The community was giving up land and, in the case of Lolkisale, a significant share of its income from Treetops. In return, the community expected well-managed grazing banks, dedicated crop protection teams, and a share of the tourism profits. From Honeyguide's perspective, these were reasonable expectations, and it has worked to ensure that the community, as the shareholders, reap these returns from their community-based enterprise.

While Honeyguide recognized the value of prioritizing community livelihoods, both in terms of social outcomes and the ecological imperative to preserve wildlife habitat outside Tarangire NP, convincing donors of this was more challenging. The World Wildlife Fund and large international wildlife conservation donors are still largely convinced that heavy-handed anti-poaching strategies are the best way to protect wildlife on community land. This translates into a funding bias toward anti-poaching initiatives. Honeyguide was able to secure donor funds to commence anti-poaching operations in Randilen WMA, drawing from its track record in Enduimet, Burunge, and Manyara Ranch, but realized quite quickly that if the community did not want poaching on their land, then it did not occur. Innovative strategies for tracking poaching activities, including canine units, undercover informants in communities, and a modest squad of village game scouts (VGS), were sufficient for essentially eliminating elephant poaching in Randilen. While ivory poaching had been a concern in 2012–2013 when Honeyguide first started working in the area, I was surprised to find that zero elephants had been killed in Randilen over the past five years at the time my fieldwork in 2019–2020. Honeyguide's effective anti-poaching strategy was partially to thank for this, but it was also largely a reflection of community members buying into conservation through a give-and-take reciprocal relationship with the WMA. Through discussions with community representatives, Honeyguide realized very early on that carrying out anti-poaching patrols in the wet season was a non-starter across Arusha member villages, as this was the time of year when maize was grown. The community had no interest in being policed in those months when they were so desperately concerned about herds of elephants in their farms on a nightly basis. Honeyguide listened and, in a gesture of good faith, stationed one of the VGS patrol vehicles in Nafco and the other in Naitolia to help people direct elephants back into the reserve area at night. Unlike other nongovernmental organizations (NGOs) and tour companies that brand their vehicles with their logos, Honeyguide was adamant that it wanted the vehicles to bear no company symbols so that the community associated them directly with Randilen WMA. The hope was that people would realize that it was *their* WMA that was carrying out activities that they valued and viewed as centrally important to their livelihoods. Community members very much appreciate these efforts. As one Arusha man (about sixty-five years old) from Naitolia explained, "There was a time when we did not like the WMA because we thought that they were the ones bringing these wild animals inside our villages. But we like it now because when the elephants raid our farms, they come to help."

I carried out emplaced fieldwork in Nafco village in April–June 2020 during the harvest season and participated in some of these evening expeditions into people's maize fields to wrangle elephants back into the WMA. If one thinks herding cattle is challenging, shepherding wild savanna elephants while driving through maize fields is sheer madness. On some evenings, I counted herds of sixty or seventy elephants in people's farms, which I was shocked to learn could easily destroy an entire five-acre maize field in a single night. What I was perhaps most struck by was the amount of damage that the vehicles themselves do to people's farms in pursuit of the elephants. Each night, a few Arusha or Kisongo passengers accompanied our team as we screamed and yelled and fought what felt to be a never-ending battle to steer the elephants back into the WMA. When I asked the VGS about the impacts

Honeyguide and Randilen Wildlife Management Area staff help a woman in Nafco village shuck maize in the harvest season. Small gestures like this help foster local support for conservation by showing community members that Randilen's management team cares about people's livelihoods. Photo by author in 2020.

of vehicles on people's crops as compared to the elephants, they replied that if the elephants were left undisturbed, the damage would be far greater, an assessment which I agreed with. But what was more telling was the response given to me by one of the Arusha cultivators who hopped on board: "We are just happy that they are here to help." I realized quite clearly in that moment that what was significant to those farmers was not, in and of itself, the economic impacts of the crop-raiding reduction efforts, but the *gesture* that symbolized to people the fact that conservation was not just about protecting wildlife at the expense of people. It was also about protecting the things that mattered to community members. In this case, it was about protecting people's crops. When we returned to the small roadside restaurant in Nafco each night after hours of exhausting efforts chasing elephants from fields, ordinary villagers took turns buying the dinners of the VGS—small gestures in return to show them how much the community cared about the work they were doing.

In Damian's view, communication between the WMA and the community is "probably the most important part of the work" that Honeyguide is doing and is "probably the most important part of community-based conservation" in general. Damian likens the WMA to a public liability company that is trying to attract its shareholders to continually invest in it. Essentially, its business model is based on the community continuing to invest its land and natural resources. To win these continued investments, the WMA's potential returns must constantly be communicated back to the community. Damian contends that WMAs should prioritize articulating to communities the benefits of investing village resources, just as NGOs convey to their donors, through visually appealing annual reports, the impact of their funding initiatives. Big international NGOs regularly spend thousands of dollars on graphic design to publish aesthetically pleasing pamphlets and brochures because they know that it is crucial for their ability to secure donor funding (Igoe 2017). And yet, as Damian points out, conservationists often assume that communities will willingly invest in WMAs without any effort taken to communicate back to them the returns on their investments. Despite its importance for community-based conservation, donors have little interest in funding communication efforts between WMAs and communities. They would rather invest money to "catch poachers," without recognition of the fact that by building community support for conservation, the need for anti-poaching declines in tandem. This highlights a potential point of intervention in the field of conservation. Rather than defaulting to the "fines and fences" model of top-down conservation enforcement, an empathetic approach prioritizes open lines of communication between communities and conservation managers and active listening to people's livelihood concerns. Like a company that must remain answerable to its shareholders, conservation areas must remain accountable to their local communities to ensure that the costs and benefits of conservation practice are equitably distributed.

A Theory of Change

Honeyguide focused its management efforts on generating community support for the WMA, which it viewed as centrally important to Randilen's long-term success. These attempts were informed by Honeyguide's "theory of change." As described in Honeyguide's 2017–2021 strategic plan, "Honeyguide works to bring about behavioral changes at the community level that result in communities protecting wildlife and rangeland habitats, based on their own social, economic, and cultural interests and values" (Honeyguide 2017, 1). Put in different terms, Honeyguide was working to achieve what Agrawal (2005) refered to as environmentality, whereby the community comes to develop positive attitudes toward the WMA in part due to its changing institutional dynamics. My research shows this to be occurring. Rather than frame this process in terms of government and subjectivity in a Foucauldian fashion, however, I am inclined to interpret this case in terms of the community's growing appreciation for Randilen's tangible contributions to local livelihoods and well-being. I consider Honeyguide's efforts as an example of empathetic conservation characterized by an overarching commitment to learning the interests, worries, and aspirations of community members and integrating these perspectives into the design of the conservation model.

A key aspect of Honeyguide's attempts to engender attitudinal and behavioral change was about building up people's trust in the WMA as a "friend" rather than a "foe" (Wright 2019). Honeyguide provided much-needed management capacity for the WMA and tailored Randilen toward protecting, rather than policing, local agropastoral livelihoods. In the words of a female Kisongo elder (about eighty-five years old) from Lemooti, "Randilen WMA protects us. The VGS protect us from wildlife and guard our community land against outside interests and farmers. The WMA also conserves our pasture during the rainy season so that when there is a drought, we have a place to graze our livestock." One management strategy that worked especially well in this regard was the enforcement of the elite outsiders who had gained access to village land prior to the establishment of the WMA. Keeping out the cattle of powerful politicians was no simple task. In the early days of WMA enforcement, VGS sometimes confiscated trespassing cattle led by local herdsmen on behalf of absentee owners in Arusha town, only to receive phone calls from regional police officers a few hours later asking for the herds to be released without fine. The cattle barons, of course, were well connected politically, making the enforcement of restrictions at their expense challenging. Nonetheless, through dedicated support from Honeyguide, Randilen VGS learned to hold their ground in these instances by continuing to subject the encroachers to standard fines prior to releasing the cattle. On one occasion, a wealthy herdowner drove in from Arusha town and led his sizeable herd into the center of the WMA's reserve area in a belligerent act of protest. He was swiftly confronted by armed VGS in green uniforms, escorted out of the area, and fined for his transgression. With time, these types of altercations began to dwindle, likely because the heightened management presence in the WMA made incursions not worth the hassle relative to other areas with less enforcement on the ground. Whereas in the past, the village chair of Naitolia could be incentivized with money to allow politicians to graze their cattle in the area, the Honeyguide-trained VGS were paid fair salaries and enforced the WMA's management plan irrespective of how wealthy or politically well-connected the cattle owners were. Kisongo elites from Lemooti, by contrast, were entitled to graze their cattle in the area because they were part of the Randilen community.

Commercial farmers were also not exempt from the crackdown on outsiders utilizing village land for private gain. The areas occupied by large commercial farms in Mswakini that extended significantly into the WMA were reallocated as communal grazing areas and reserve land. Those farms had been held by outsiders who had negotiated with village councils or purchased land from villagers. As to be expected, the farmers were angry that their productive assets had been confiscated without adequate compensation and proceeded to sue the WMA in the High Court of Tanzania. Quite interestingly, the courts sided with the WMA over the farmers, and all cases in Mswakini and Naitolia that I am aware of were dropped. When I discussed this development with the WMA manager, Meshurie, we reasoned together that the state may have had an incentive to side with the WMA because it generated revenue for the central government via the Tanzania Wildlife Management Authority (TAWA) as compared to the commercial farms, which mainly produced beans and sunflowers with varying degrees of productivity. Several court cases regarding commercial farms in Lemooti, however, were still ongoing at the time of fieldwork.

Residents of Randilen's member villages began to realize during this period that the WMA was restricting access to village land by outsiders without infringing significantly on the tenure rights and livelihoods of people who actually resided in the villages. This realization was particularly significant in the Arusha-inhabited villages of Naitolia and Mswakini, where people had been fearful of eviction at the time of Loveless's (2014) fieldwork. As expressed by an Arusha man in Naitolia (about thirty-four years old), "We did not like the WMA at first because we thought it was going to be like Tarangire and prevent us from entering or grazing our livestock there. But once we realized that it was still our land, and we had the right to graze our livestock there, we started to like it." Far from a multiple land use model like the Ngorongoro Conservation Area (NCA) that restricted subsistence cultivation, a national park that displaced them, or a wildlife corridor that undermined their tenure rights, the Arusha in Naitolia and Mswakini began to see the WMA as an institution that actually served their long-term interests by helping them solidify their own claims to land. As described by one Arusha man (about fifty-four years old) from Naitolia, "There were conflicts when the WMA first started because when it was introduced to the village, some of the villagers did not know that the WMA had benefits to them. After they came back and educated people about the aims of the WMA, people came to realize that this is actually very important to our well-being. Now, almost everyone here—like 90 percent of people—like the WMA."

In helping to sway community sentiment, Honeyguide and the newly hired Randilen management team specifically targeted the village chairs who had initially opposed to the WMA due to the influence of outside elites. Honeyguide helped foster direct dialogue with those village chairs and WMA staff and integrated them into important governance discussions. One Arusha man (about forty-five years old) in Naitolia laid out the implications of these efforts in an interview in 2019:

> The interesting thing is those who were initially against the WMA are now the ones operating it. They were hired as VGS to oversee the environment of the WMA and enforce the seasonal restrictions on livestock grazing. Some are also part of its environmental protection committee. Those people started to see the benefits of the WMA. Some of the sub-village chairs were the ones mobilizing the village to oppose the WMA. But they have been selected as AA members and they are now the ones spreading the good ideas about the WMA because they attend the AA meetings and now know the important role of the WMA for the community.

Some village chairs, however, were still opposed to the WMA in Honeyguide's early days. The village chair of Naitolia, for instance, was so against the WMA that Honeyguide simply replied in kind by withdrawing its crop protection services from Naitolia, while offering continued support for the other member villages. A short time later, residents of Naitolia began to plead with their chair to rectify his relationship with the WMA so that they could also receive crop protection. Eventually, the chair had to reconcile his relationship with the WMA as the complaints from the assembly piled up. Honeyguide recommenced crop protection services in Naitolia shortly thereafter, which local Arusha greatly appreciated. While there was a vocal minority that still attempted to "stir the pot" to destabilize the WMA in the first few years after Honeyguide's arrival, public opinion had already started shifting by 2016. Damian described to me one particularly significant event that occurred during a village assembly meeting in Mswakini around that time. In the meeting, the village chair openly preached to the villagers about the costs of the WMA. In response, one Arusha woman stood up and yelled, "You have a car. I don't see you out at night protecting us from the elephants! The WMA helps us. You do not!" The village chair proceeded to sit down quietly "with tail between his legs." In Damian's view, the woman's statement was a clear sign that community attitudes had swung in support of the WMA and that the village chairs needed to change course or lose favor entirely. Through ongoing targeted efforts, the village chairs in Naitolia and Mswakini began to realize that the WMA could bring considerable benefits to their villages through ecotourism revenue, construction of schools, and crop protection. As time passed, the same people who had previously attempted to incite opposition to the WMA on behalf of their elite financiers instead began to preach support for the WMA within the community. When I interviewed the former village chairs of Naitolia, Mswakini Juu, and Mswakini Chini in 2019, all of them voiced appreciation for the WMA and its many benefits for their assemblies. One former sub-village chair in Naitolia also contextualized his change of heart:

> When the WMA was first beginning, the village chair and us sub-village chairs were being influenced by some rich people to gain access to village land. Those people were pushing us to protest and spread opposition against the WMA. But those individuals have since been removed from the picture and we now see the good things the WMA is doing for the community. I came to see the help the WMA was bringing to my village. So, I started at that point to spread good ideas about the WMA. Before, I was spreading bad ideas, but now I am spreading good ones.

Discursive shifts were also evident in Makuyuni village during interviews in 2019 and 2020. Residents of Makuyuni reflected on their village's decision not to join the WMA, and some expressed a strong desire to become a member village of Randilen. As voiced by one well-respected Arusha elder:

> The WMA came with a bad looking face and we worried that it was going to grab our land. Its appearance has changed since then because we see that they learned some techniques for managing land in collaboration with the community. We can see that herders are allowed to graze their livestock in Randilen and it works with the community, so the WMA applies pastoral knowledge and its member villages are benefiting. I was opposed to the WMA when it was first proposed, but now I see that the WMA helps protect community land for livestock and from outsiders who want to grab it. So we wish that we had better education about the WMA at the beginning and we would not have opposed it. The negative stories we were hearing about the WMA were wrong. Now we need to join Randilen WMA to protect our land from being grabbed. We have been trying to join it since last year [2018]. We have already written a letter and taken it to the WMA. I met with the manager and Honeyguide and I told them they should take the letter to the district level, but we have not heard back.

This narrative segment from an elder in Makuyuni shows a change in perspective toward the WMA, contextualized in terms of community concerns about land tenure security and pastoral livelihoods. The elder's fear of dispossession in Makuyuni has since given way to a growing realization that WMAs may well represent the most effective tool for reducing the threat of land grabs by formalizing communal pasture in wildlife dispersal areas. Importantly, this elder's statement should not be taken as wholly representative of the entire village of Makuyuni. Some other interviewees in Makuyuni still expressed hesitancy about potentially joining the WMA in 2019, so I surveyed a randomly selected sample of village residents in 2020 to assess the generalizability of the elder's views. The results were highly mixed: just over half wanted Makuyuni to join the WMA (56%; $n=65$), while the rest did not (44%; $n=52$). Nonetheless, it is safe to say that a significant number of people in Makuyuni now support joining the WMA, a marked shift since the village initially refused to become part of Randilen.

Circling back to the role of Honeyguide in the development of Randilen, its goals were threefold: First, it sought to build up the WMA to a point where it did not require an external donor to support its operations, ideally within five years. Second, it wanted to cultivate community support for the WMA by ensuring that people genuinely valued its contributions to their lives. Third, it worked to establish a business framework that safeguarded wildlife habitat. Poaching was a minor threat to this third goal compared to the conversion of rangelands to farms. In working toward these three interrelated goals, Honeyguide came to find that what mattered most to the community was self-determination and livelihood security. As Damian phrased the question from the perspective of community members, "Are we the captain of our ship?" People wanted to know that they were ultimately in control and could ensure that the WMA delivered things that they valued. By way of comparison with Burunge, where Chem Chem Lodge has dictated terms of the WMA, in Randilen, the community has been able to continually orient the WMA toward the things that matter most to them. Fundamental to this has been the willingness of Randilen's management team to adopt a community point of view and prioritize local agropastoral livelihoods as essential dimensions of wildlife management. Unlike the big international NGOs that often project their predetermined standards onto WMAs and communities, Honeyguide has been flexible in "observing the standards that the community wanted" and in helping to nurture the community's relationship with the WMA. Here again, I would refer to this as a process of empathetic conservation.

From Customary to Formal

Since the initial zoning plan (2012–2017) was not representative of community members' lived experiences on the ground, the Authorized Association (AA) designed the second resource management plan (2018–2023) carefully under the thoughtful supervision of Honeyguide. The revised management plan was much more reflective of the concerns of the community, and the people I interviewed across all eight member villages who were familiar with it all voiced similar appreciation. My quantitative survey data was strikingly clear in this regard, as 91 percent ($n = 620$) of respondents felt that their community was well represented by the new management plan. One Kisongo man (about thirty-five years old) from Lengoolwa articulated his views during an interview in 2020:

> The first reason why some people opposed the WMA at the beginning is that there were a few individuals that came and told us that the WMA would not be good for livestock keepers. They told us that our livestock would not step foot on that land. They were not giving us the real information about how the WMA would operate. They said that WMAs are just parks for keeping wildlife. For us Maasai, we value our livestock like we value our own lives. Our farms are not as important to us as our livestock. Our WMA has a management plan that allows livestock to graze based on seasons. There is a yellow zone where we are allowed to graze each year in the dry season. They usually open it up for our livestock in July. And then there is the green zone that they keep aside unless there is a very bad drought. If there is a drought, that area is kept for us to use. So, we love the WMA management model because it supports our livestock.

As this man's narrative makes clear, misinformation about the WMA was rampant when it was first being established. Hurried and top-down zoning processes led by AWF exacerbated worries that livestock keepers were going to be dispossessed of land. Herders have since come to realize, however, that the new zoning management plan is a means of formalizing a rotational livestock grazing system that takes into account seasonal variations in rainfall. As one Arusha man (about forty years old) from Mswakini Chini described to me during an interview, Honeyguide made a strong effort to take stock of community livelihood priorities across all the member villages before helping to formalize the new plan. In his words:

> They did a survey, as you are doing now with this interview. But they used survey forms. They went to all the villages and collected people's opinions

and views. Then they made a report. Then they took the report to each village office, and in the village assembly meetings, the villagers came and read out the rules and said these are the rules that we came up with ourselves. Is there anything to add? Is there anything to remove? And so we were the ones who agreed in such a way.... So in my view, it is a successful project because we are benefiting from the grazing plans.

Taking a genuinely participatory and community-based approach to devising the new zoning plan ensured that WMA management represented the interests of the resource users on the ground. As one Arusha woman (about forty-five years old) from Mswakini Chini put it, "I like the seasonal grazing restrictions because when the grass finishes in the village, we are allowed to graze our livestock in the grazing area in the WMA. We know that in times of need there is always an area reserved for us—our reserve grass."

One interviewee in Mswakini Juu, however, did express some concerns about boundaries between the WMA, Tarangire NP, and the village in 2019, suggesting that some disputes had not yet been fully addressed. However, during follow-up conversations in 2022 and 2023, the interlocutor suggested that those issues had since been resolved.

The revised Management Zone Plan establishes two primary zones in the WMA based on areas that were resurveyed in April 2018 (RWMA 2018). A central Tourism and Photographic Management Zone (reserved area) covers 215.53 km^2, and three peripheral Tourism and Livestock Management Zones span 96.66 km^2. Within each area, there are lists of acceptable uses that are agreed upon by the AA and enforced by VGS. Notably, the core tourism and photographic zone overlaps the preexisting Lolkisale Conservation Area. While the map on paper appears highly legible, it is actually a representation of key landmarks on the ground that are used by the local community. Rivers, identifiable trees, and hills serve as reference markers and boundaries were agreed upon by the AA. Inside the reserved area, hunting, mining, off-road driving, tree cutting, capturing wild animals, cultivation, charcoal burning, and "unauthorized livestock grazing" are prohibited (RWMA 2018, 30). Local herders were very particular to specify that "unauthorized" grazing was not allowed (RWMA 2018, 28–29). The purpose of this provision was to ensure that the AA retained authority to decide whether grazing would be allowed on a flexible basis depending on season and variable environmental conditions. Flexible livestock grazing restrictions are enforced by the VGS at the discretion of the AA. As articulated by the WMA chair from the period of 2017–2022:

> We formed the idea of having a grazing bank because livestock is our first priority. In the early days of the WMA, our main concern was whether livestock would be allowed to graze in the WMA. Because some WMAs like Burunge do not allow grazing. Once we knew that this WMA could be used for grazing, we came to know that this WMA was a good thing because we were struggling to find places to graze our livestock. Those armed rangers that you see in the WMA are not catching our livestock and stopping them from entering the WMA. We have chosen VGS from each of the member villages. They are not government rangers. People were chosen from the villages because they are pastoralists. We would never trust someone who is not a pastoralist to come and protect the WMA because they might sell the grass for other purposes. But if you put pastoralists in charge of management, which we have, then they will take care of the place in a way that supports the Maasai community.

The notion held by some conservationists that herders are universally opposed to restrictions on livestock grazing is not true, as there have historically been numerous customary institutions and normative practices for regulating access to pastures at particular times. Pastoralists, however, take issue when restrictions are imposed on them from outsiders in top-down ways that are fixed across space and time, and based on a different way of conceptualizing and managing rangelands (Bluwstein 2019). In the context of Randilen WMA, the vast majority of people I interviewed across all eight member villages took no issue with the general restriction on grazing inside the core photographic and tourism zone because they were happy to keep it as a buffer for times of dire need. In the words of a Kisongo man (about fifty-three years old) from Lemooti:

> The WMA is good because it helps us in several ways. First of all, it conserves the area and stops the rangelands from being converted and developed. Secondly, it keeps the land for us to graze our livestock there seasonally. There is one zone where our livestock can graze all year, but we only allow livestock to graze in the protected zone when they do not have enough food outside. In those cases, we speak with the AA representatives, and we are then given a part inside the protected zone to graze. So the reserved area is like food storage for our livestock.

As this man's narrative segment reveals, local herders appreciate the way that the WMA's zoning plan helps them conserve grass for the dry season. Arusha livestock keepers voiced similar sentiments on the zoning plan. As one Arusha man (about thirty-eight years old) from Mswakini Chini described during an interview:

> Before Randilen, we did not have grazing plans. The WMA helped us implement them. In the WMA, there are some areas where livestock are not allowed to reach. But we like those restrictions on grazing because we are the ones who made the rules, and we are following them. Of course, some people occasionally break the rules and try to graze in the places where livestock are not allowed to reach. But if we are the ones breaking those rules, we are caught, and we are fined. *We* are the ones who made the laws, so if we break the laws, we pay the fines.

Interviews made clear that local livestock keepers consider the new grazing plan to be equitable and representative of community livelihood concerns. For further quantitative contextualization, 89 percent ($n=607$) of people surveyed viewed the WMA's environmental protection model as a form of community-based conservation, and only 10 percent ($n=71$) considered WMA restrictions to be indicative of fortress conservation. The crux of the matter is that herders view the spatial and temporal restrictions on livestock grazing as *serving* the interests of the livestock economy, rather than undermining them.

According to the official management plan, livestock grazing is only permitted inside the photographic zone during "extreme weather conditions" and subject to permits by the AA (RWMA 2018, 30). Essentially, then, the core photographic zone serves as a "drought reserve," which is a traditional Maasai technique for managing pastures in semi-arid environments (Goldman 2011, 73). In 2022, for instance, a severe drought in northern Tanzania necessitated that the photographic zone be opened to local pastoralists as needed for livestock grazing. Traditionally, Maasai elders oversee the enforcement of drought reserves, but in Randilen, it takes place through the formal institution of the AA. This is a very significant development because, as McCabe et al. (2020) point out, traditional customary institutions among the Maasai for managing rangelands—through territorial sections, clans, and the age-set system—are being superseded in contemporary Tanzania by the formal institution of the village. The emergence of the village as a political tool for regulating access to pastures can help defend pastoral land from state-led grabs (chapter 1), but it can also undermine the management of semi-arid rangelands by parceling environments with unevenly distributed sources of water and productive grasses into small areas that are managed by independent parties. While the WMA is also a formal institution that has the potential to undermine traditional grazing patterns, in Randilen, it is being wielded by the Maasai as a strategy for ensuring a collectively managed dry season grazing bank. If I were to distill this entire book down to the single most significant detail, I believe it to be this: the WMA fits within a traditional cultural framework for managing semi-arid rangelands and promotes resilience in the face of uncertain and variable environmental conditions. From the perspectives of local herders, this is the most significant reason why they have come to appreciate the area, and this is where I see much common ground with conservationists. From a conservation perspective, the photographic zone protects wildlife habitat in an important elephant dispersal area from competing land uses. From a pastoral perspective, the photographic zone protects reserve grass from cultivators, land grabbers, and livestock keepers who do not respect collective institutions.

When I posed the question of how people feel about the WMA during interviews in 2019–2020 across Randilen's member villages, the most frequent response put forth was, "I like the WMA because we use it for livestock grazing and it has good grass." The important context here is that most households depend on the WMA for pasture—82 percent ($n = 556$) of survey respondents reported grazing their livestock inside the WMA. As one Arusha man (about fifty years old) in Naitolia said, "We know that, if there is a major drought, we will be able to graze our livestock in the whole area, so we are happy to have the WMA." Interviewees were also particularly appreciative of the fact that politicians can no longer graze their livestock inside the WMA by negotiating in nefarious ways with village councils. As a female Kisongo elder from Lemooti (about seventy-five years old) put it, "The WMA is our grazing bank. We are grateful that it keeps our livestock healthy in times of drought and now there are no livestock coming from outside into our area." VGS have vehicles and weapons and are able to patrol and enforce the area on behalf of the community. Since the Maasai have traditionally valued reciprocity across regions, the current management model of the WMA stipulates that the AA retains the authority to negotiate livestock grazing access and provide permits to people as needed. The AA is thus able to permit requests from Maasai herders from afar in exceptional circumstances. Generally, however, livestock from outside the community are not permitted inside the WMA's community grazing area. This insulates the community from the rich politicians who have historically gained access to the area through the Naitolia village council.

A key aspect of the revised zoning plan is that some villages also have areas that are classified as mixed-use zones. In these areas, community livestock grazing is permitted, while livestock from outside the community are prohibited from entering without permission. Livestock keepers are also allowed to establish *ronjo* bomas in these areas. Part of the reason for the initial protests in 2014 was that local herders had heard that VGS were in the process of burning unauthorized *ronjo* bomas, which caused a very strong visceral reaction among people who had livestock at stake. While the revised management zoning plan includes upper limits on the number of livestock allowed in the mixed-use areas, these limits were determined by the community and are much higher than the villages are able to reach. Cattle are limited to 100,000 heads, goats are capped at 80,000, and sheep at 50,000 (RWMA 2018). For context, Lemooti had approximately 20,500 cattle total in 2018, the highest of any member village despite its small human population size (RWMA 2018). Naitolia had the second highest number at approximately 5,500, with the remaining villages having less than 5,000. This means that the maximum number of cattle in the entire WMA is approximately 50,000–55,000, keeping it well under the upper cap. The community intentionally selected upper limits that were great enough that they would not lead to conflicts in practice because the sociocultural and economic value of accumulating livestock still endures in Maasai society. Livestock numbers tend to reach natural caps relative to changing environmental conditions, and community members are not worried about exceeding these upper limits.

Other resource uses in the mixed livestock zones are regulated in various ways. Harvesting plants for traditional medicines must be authorized by the AA, but trees cannot be ring-debarked or uprooted. Collection of dead firewood is permitted, but live tree falling is prohibited, as is charcoal burning. The restrictions on charcoal production are not specific to the WMA since the practice is illegal throughout Monduli district. During interviews with District Forest Officers, it was revealed that charcoal burning is an ongoing problem in the area, particularly in the Arusha dominated villages. On several occasions during my fieldwork, I encountered people burning charcoal openly in their homesteads, or doing so more discreetly at night. District Forest Officers take charcoal production quite seriously and generally issue stern fines to those they catch, as they view it as a primary driver of deforestation in the village areas around Randilen WMA. Within the mixed-use zones, other land uses like hunting, cultivation, and mining are prohibited. Beekeeping is permitted in both zones, with an upper limit of ten thousand bee hives in both areas, which the community is also unlikely to reach.

Restrictions on cultivation were welcomed on the Lolkisale side of the WMA by the pastoral Kisongo but were a topic of significant concern for the Arusha on the Naitolia and Mswakini side. Given the importance of farming to the Arusha mode of production and way of life, local Arusha were concerned that the initial zoning plan restricted their abilities to cultivate too significantly. The revised zoning plan demarcated a modest zone that would be kept aside for livestock keeping and tourism, while leaving the majority of the villages open for cultivation. The mixed livestock and tourism zones in Mswakini Chini and Juu are thus quite small compared to the one in Lemooti. Importantly, a binary opposition between farmers and herders does not exist in this ethnographic context, even despite ethnic differences in the economies of the Kisongo and the Arusha. Though they consider themselves to be farmers, the Arusha greatly value the livestock grazing banks in their villages. Their move from the Meru highlands to the Monduli lowlands has meant that livestock keeping has become a highly productive way to use land. Livestock is especially valuable to the Arusha since their economy and social life has become entangled with the Kisongo, with whom they participate in shared rituals, exchanges, and marriages, and with whom they trade in grain, livestock, and labor. Small stock serve as liquid cash accounts that beat inflation and can be brought to market anytime. Cattle hold significant exchange value with the Kisongo and are still the de facto currency of Maasai social life. So, even though the Arusha favor their farms much more than the Kisongo in terms of investing labor and physically tilling land, they also care deeply about their livestock, and thus appreciate the grazing banks that the revised resource management plan provides them.

King of the Hill

Randilen WMA has become a formal institution that local herders use to manage rangelands, but "institutional layering" still complicates the political landscape of the area (Lesorogol 2022). Pastoralists from the five Lolkisale-side villages of the WMA have historically relied on Lemooti as a crucial seasonal grazing area, given the alienation of pastoral land in the Lolkisale area for commercial farming. Access to this area is now technically governed by the AA and enforced by VGS as part of Randilen WMA's mixed livestock and tourism area. While the area continues to provide a crucial grazing bank for pastoralists coming from Lolkisale, an issue developed in 2019–2020 due to overlapping jurisdictions of the village and that of the WMA. Both are formal institutions for managing land that are supported by state law (see McCabe et al. 2020), but the question of which should supersede the other is sometimes contested.

Prior to the 2019 elections, there was no real issue as pastoralists from the Lolkiale side entered the area as needed, established temporary *ronjo* bomas to tend to their stock, and abided by Randilen's generous upper limits for livestock. In 2019, however, the young man who had struck it rich in the tanzanite trade (chapter 4), and who was initially opposed to the WMA, decided to run for office. I stayed in Lemooti and carried out interviews across its sub-villages in the months leading up to the election, and I was hard pressed to find many people who supported his claim to the village chair position, at least openly. Most were instead wary of what would happen if someone with so much "money power" were to take up the position, knowing that village councils have great authority to allocate land within their territories. By then, I had developed a sense that the WMA served community interests, and I was concerned about how this individual, who did not seem to care much for collective institutions, would affect the sociopolitical landscape of the area. Unlike other people in Lemooti, the *tajiri* ("rich person" as he is referred to locally) lives atop a large hill overlooking the WMA. He has built a large house with a high tower atop, not unlike a castle. It looks rather out of place in the village, but I could not help but think enviously that the view from his tower must be stunning. When I interviewed some of his neighbors about how they felt about him building a castle atop the hill, and about what he was like as a person, one man angrily explained, "This man has no respect. He brings thousands of cattle through here whenever he wants. They trample my maize farms and he does not even apologize." I had thought at the time that there was no way the *tajiri* would succeed in his bid for office with that much dislike from other community members. Evidently, there was more going on. A few months later, I was staying in Lolkisale-proper carrying out interviews, and I heard a group of people in the street worriedly discussing the election result in Lemooti: the *tajiri* had won.

I was shocked at first, and when I returned to Lemooti to carry out follow up interviews with people, some still insisted that they had not supported his bid, but a notable number had shifted their views. When I poked to find out why, it became evident that there was money involved in swaying votes. As one interviewee explained, "He used his money power to become the chair," implying that he had bought votes. Knowing that he was one of the vocal minority who was opposed to establishing the WMA in the first place, and that Lemooti had given the largest portion of land to the WMA, there was a sense in the greater Lolkisale community that the *tajiri mwenye kiti* (rich village chair) might be scheming about something. Sure enough, immediately after taking power, he began to reallocate the temporary *ronjo* bomas of livestock keepers from outside Lemooti to herders from his village. While he did not have the physical means to "block" outside pastoralists from entering the area, he signaled to the incoming herders that they must first register with the village office before entering the area. Preventing them from setting up their own temporary *ronjo* bomas in the mixed livestock area made it difficult for livestock keepers from the greater Lolkisale area to graze their livestock there. As one Kisongo woman (about sixty years old) from Lengoolwa exclaimed during an interview in 2020, "The new chair of Lemooti is becoming a big problem because he is trying to chase us away from Randilen's communal grazing areas!" A male elder in Oldonyo described these dynamics in detail:

> Nowadays, the new village chair of Lemooti is trying to block us from accessing our grazing areas over there. Before he was in power, every herder could have a place over there. Lemooti was a village for grazing. But nowadays if you visit after being away for two months, you may find that your *ronjo* grazing area has been allocated to a person living in Lemooti. But for the Maasai we share grazing areas. So, with this new rich village chair, we have this feeling that we will not be allowed to graze beyond the river in Lemooti anymore. This is what we are starting to see because he is trying to eliminate us from the area.

The *tajiri mwenye kiti* was operating under the assumption that the Village Land Act trumped the revised WMA regulations in terms of land tenure and rights. Harking back to the discussion in chapter 1, Maasai in Tanzania have wrestled for the past twenty-five years with the question of which institution is more stable and enduring for managing pastoral rangelands—villages or WMAs. Technically, in Lemooti, the areas overlap, as the grazing area in question was within the WMA *and* village land. Much like the Game Controlled Area controversies of the past, the crucial question in this context is, When there are conflicting laws governing WMAs and villages, which one takes priority? Since there are three types of land in Tanzania (general, village, and reserve land), WMAs make it somewhat unclear which category these overlapping areas fall under. Technically, Lemooti is still village land, but it is also a reserved area within the framework of the WMA. While some lawyers have suggested that the Village Land Act could potentially trump the WMA regulations in court, my own interviews with government officials have suggested that the WMA institution now formally supersedes individual villages. This means that the AA has the ultimate authority to determine access to, and use of, land in the mixed livestock and tourism area, not the Lemooti village council. Despite initial fear from Maasai leaders in Loliondo (chapter 1) that

WMAs would not offer permanent tenure status on par with villages, land that is classified as a WMA with an agreed-upon and ratified land use plan is formally protected by state land and wildlife laws and is as secure as tenure can be in Tanzania.

The *tajiri mwenye kiti* of Lemooti is someone who I would call a "villagist"—that is, a pastoralist who supports and values the formal institution of the village for governing access to land rather than communal forms of tenure that involve reciprocal access rights across semi-arid rangelands (cf. McCabe et al. 2020). During my conversations with people in Lengoolwa, Nafco, Oldonyo, and Lolkisale-proper following the elections, the new *tajiri mwenye kiti* in Lemooti was consistently mentioned as a concern for pastoral livelihoods. As one Kisongo man (about sixty-five years old) from the outskirts of Nafco said during an interview, "The rich village chair has already started blocking our livestock from grazing there. But where else can we go? That is our only grazing area." As another Kisongo man (about fifty years old) from Oldonyo lamented, "He does not care about us, he wants only to control access to the grazing area in that territory." Pastoralists from Lolkisale were aware that the *tajiri* had a large herd of cattle and viewed his new position as a means of consolidating power. Interviewees interpreted his actions as attempts to secure his own private grazing area at the expense of pastoralists from other villages who had smaller herds and less political clout.

Concerned about the potential for this fellow to undermine community support for Randilen, I raised the issue with the WMA chair. Meshurie explained to me that with someone like the *tajiri mwenye kiti*, one has to be diplomatic. Meshurie's philosophy was that it was best to carefully bring the *tajiri* into WMA governance meetings to make him realize that there was much to be gained from building solidarity and acting in good faith. This, however, proved challenging in the early days of the *tajiri*'s tenure. During the WMA governance and equity workshop in Mto wa Mbu in 2019 (chapter 5), not long after the local elections, the *tajiri* was notably absent. I asked Damian of Honeyguide at that meeting what he thought about the *tajiri*'s rise to power, and whether it jeopardized the collective institutions of the WMA. Damian's response was that since the WMA had already been formalized, it would be difficult for a single village chair to oppose it at this stage ("it's like a machine that's already in motion"). Indeed, the WMA institutions are supported by state law, making them difficult to overrule once they are established. However, resistance is always a possibility if people do not view laws as legitimate. On one occasion in early 2020, for instance, the *tajiri* refused to sign off on the WMA's routine distributions to each member village, arguing that Lemooti deserved a larger piece of tourism revenue because it had given up more of its land to the WMA.

To avoid speculating about the *tajiri* based only on secondhand information, I thought it prudent to interview him myself to develop a sense of what his thoughts were on collective solidarity and the WMA as an institution. Unfortunately, he proved very challenging to track down. He was rarely in Lemooti and was often at the mines in Mererani or the tanzanite market in Arusha town. Finally, in May 2020, we both attended a Randilen WMA AA meeting together in Makuyuni, and I had a chance to observe the nature of his participation closely. As the AA members and government officials emphatically clapped their hands in support of the WMA and chanted "*tembo, na maendelo; maendeleo, na tembo*," I watched him as he hesitantly began to join in. He was surrounded by unanimous support for the WMA, and I am sure he began to feel part of the Randilen community. By the end of the meeting, he was making constructive suggestions as Meshurie took extra efforts to engage him with questions and make notes on a flipboard based on his suggestions to ensure he felt listened to. I could sense that some of the *tajiri*'s grievances with the WMA were starting to dissipate now that he had a seat at the table and felt included in the WMA's governance dynamics. While it was far too soon to ascertain whether his constructive engagement with the WMA would be enduring, I could not help but wonder whether the *tajiri* symbolized the last chip to fall in place in terms of community support for the WMA. He may well have been holding some of his cards close to his chest, but he appeared to be earnestly engaged. I suspect that he was surprised by how positive everyone else was about the WMA and how welcoming his peers were to him, even knowing that he had opposed the WMA openly in the past.

After touching base and exchanging numbers at the meeting, I finally tracked the *tajiri* down for a formal interview in June 2020. We met in the backseat of an all-black BMW with tinted windows in the parking lot of the tanzanite market in Arusha. When I asked him directly about the WMA, he explained that he had come to see its value for the community, but he still felt that the village institution should trump the WMA. He explained diplomatically that the village need not entail universal prohibitions on incoming pastoralists, but that it should be respected enough that herders from elsewhere register at the office "and follow the correct procedures." In his mind, the village was still the dominant land management institution that should be upheld. While it was difficult to disaggregate the *tajiri*'s personal stake in the matter from his philosophical stance on rangeland management institutions, his interest in preserving the political self-determination of villages was a fair one—institutional interplays between land and wildlife management policies in WMA member villages must be thoughtfully addressed if WMAs are to be considered socially successful models of conservation.

Concerns about WMAs superseding the tenure status of Maasai villages speak to the worries of rich herders with sizeable stakes in range management dynamics, but also to deeply entrenched fears about evictions and dispossession given wider trends of fortress conservation in Tanzania. Maasai communities consider villages known entities with the potential to secure community land, while WMAs are still viewed in some circles with skepticism and apprehension. Kisongo leaders in Randilen's member villages spoke directly to these initial worries and their changing feelings about the WMA during interviews in 2019–2020. As a well-respected male Kisongo elder (about ninety years old) from Lemooti shared during an interview in 2020:

> There was a time at the beginning when we were apprehensive about establishing a WMA. We feared that the government would bring soldiers and displace us from our homes. But that did not happen and we came to see that the WMA helps our community a lot with grazing access and community development projects. We no longer fear the WMA because everything is open. If we have a meeting, we just call them and we speak to them and if they have a meeting, they call the village, and speak to the village. We communicate well so we are working together with the WMA.

A male Kisongo traditional leader (*olaunoni*) (about thirty-six years old) from Lengoolwa echoed similar sentiments:

> There is no problem with the WMA because we are the ones who decided for it to be here. We have a good relationship with the WMA. Yes, there was a time before the WMA started when people were cautious because we did not know what will happen in the future. There was fear that the land was going to be grabbed and taken away. But we came to trust the WMA because the great fear of pastoralists is that we will not be allowed to graze our livestock in an area. That is a deep fear for us, so once we knew for certain that our livestock would be able to graze there peacefully, we came to trust the WMA.

Narratives from Maasai leaders in Randilen's member villages consistently reveal a transition from feelings of apprehension to a sense of trust, as herders realized the potential for the WMA to secure pastoral tenure rather than undermine it. Randilen has continued to cultivate these sentiments through participatory governance practices and a livelihood-based model of management. Despite the initial moves by the *tajiri mwenye kiti* to prevent herders from Lolkisale from grazing their livestock in Lemooti seasonally using the village institution, he was ultimately unable to do so as the AA upheld a reciprocal intervillage grazing plan to be enforced by VGS. Layering of the WMA over villages has thus destabilized the private stakes of elites in communal land and empowered the community to govern its common pastures in an equitable and collaborative way.

CHAPTER SEVEN

Complexities of Community-Based Conservation

Ethnographic analysis makes clear that Randilen Wildlife Management Area (WMA) is well received by its local community and plays a significant role in supporting agropastoral livelihoods. Despite this promising finding, complexities related to the WMA model are worth highlighting here if lessons from Randilen are to be effectively applied elsewhere. A significant issue to be thought through by policymakers is the effect of WMAs on the economic landscape of community-based conservation on village land.

Prior to the WMA era of community-based conservation, private investors could negotiate the terms of their contracts directly with villages. An investor operating a private business had to register with the central government and pay a certain percentage of the company's earnings in taxes submitted through the Tanzania Revenue Authority (TRA). There are numerous bureaucratic steps involved in registering a business of this nature in Tanzania, but it generally involves applying for a tax clearance certificate, a business license, a value-added tax certificate, and a taxpayer identification number. In terms of the tax structure, companies paid their licensing fees and taxes on their earnings directly to the government. Concession payments to villages came directly out of an operator's earnings, paid as a fixed amount for the concession or a percentage share (Gardner 2016). Typically, operators and villages negotiated a bed night fee that would go directly to the partnering village bank. Bed night payments were paid per guest, per night, based on a negotiated amount between investors and villages. In Lolkisale, Treetops had been paying Lolkisale village a $15 bed night fee. Following the legislative reforms and the emergence of the WMA framework for conservation on village land, the basic business model of the photographic tour operators in the Lolkisale area did not fundamentally change. They still operated as private businesses and paid licensing fees and taxes directly to the government based on their earnings.

What changed, however, was the payment system for managing bed night fees (Sulle and Banka 2017). Rather than being paid directly to a single village on a case-by-case basis, the bed night fee became a formalized tax that was centrally collected by the Tanzania Wildlife Management Authority (TAWA).

The WMA regulations of 2012 stipulate the revenue-sharing model to be applied across all WMAs in Tanzania, which are disaggregated into ones supporting photographic activities and those with trophy hunting blocks. In WMAs where hunting is administered, TAWA takes 25 percent of the block fee and 65 percent of the game fee. Since the Kisongo of Lolkisale, in collaboration with the African Wildlife Foundation, had zoned out trophy hunting, Randilen WMA was categorized as a photographic area. As things stand at the time of writing, TAWA takes 30 percent of bed night revenue, Monduli District Government takes 5 percent, and the remaining 65 percent is returned to Randilen WMA. Of this remaining 65 percent of the gross revenue, 50 percent goes back into the operational costs of the WMA, and 50 percent is distributed to member villages. In the case of Randilen, the funds that remain with the WMA are administered by the professional finance team that was put in place to manage them. The formal bed night fee in Randilen remained at $15 per night after the WMA was established.

WMAs also introduced entrance fees that mimicked those of national parks and covered a period of up to twenty-four hours. Generally speaking, the burden of these costs falls to tourists and not the operator, though lodges can assist in organizing permits. In such instances, tourists are made aware of the added fees, which increase the total price of their stay slightly but directly support the community and WMA's operations. These WMA-level fees follow the same formal tax structure as the bed night fees from the lodges. The entrance fees of Randilen, as of 2025, are $10 for a noncitizen and 10,000 Tsh ($3.81) for a citizen ($5 and 5,000 Tsh/$1.90 for children). Camping fees are 15,000 Tsh ($5.71) for a citizen and $20 for a noncitizen (5,000 Tsh/$1.90 and $20 for children). Tanzanian registered vehicles are charged 10,000 Tsh ($3.81) per day. Walking safaris can also be arranged for a price of $10 per person, and night drives cost $20 per person. Village game scouts (VGS) can also be hired as guides for $10 per day. A value-added tax that goes directly to TRA is additional to these amounts. Randilen WMA thus generates revenue, for itself and member villages, through bed night fees, entrance permits, and WMA activity fees. Accommodation fees for staying at the lodges and tented camps inside the WMA are paid by tourists directly to the tour operator. The operator must subsequently pay a $15 bed night fee to the WMA (via TAWA) since the village is no longer responsible for negotiating concessions or payments. Earnings generated by investors from tourist accommodation payments are taxed by the central government in the same fashion that they would be outside WMAs and in this sense, their business operations carry on largely independently. However, the formalization of bed night fees can be a major source of conflict in WMAs both from the perspective of villages and investors, potentially undermining community support for WMAs (Nelson et al. 2006; Sulle 2008; Sulle and Banka 2017). From the perspective of lodges, the pricing of accommodation offerings is crucial for determining economic margins, as tourist payments must cover the running costs of their lodges, including salaries for staff, maintenance fees, food, supplies, equipment, and vehicles, while still offering suitable returns on their investments. As of March 1, 2022, the current prices per night for the lodges in Randilen, according to TripAdvisor, ranged from $360 per night for Boundary Hill to $1,240 per night for Treetops, with the other lodges in a similar ballpark. Ecoscience, discussed in the following pages, is at the upper end and charges around $1,500 per night. Nimali and Kirurumu fall somewhere in the middle. From the perspective of the WMA and member villages, these prices are irrelevant, as the only aspect that affects WMA revenue is the number of beds that the lodge has, and the frequency with which it attracts guests; the bed night fee is tied only to the number of guests, so the lodge could charge $2,000 a night or $20 a night. Regardless, it would pay $15 per guest per night to the WMA as a bed night fee.

While Treetops was not directly affected by the changing WMA legislation in terms of its core operations, the revised financial structure of bed night fee collection severely affected Lolkisale's share. In 2010, following the passing of the revised Wildlife Conservation Act of 2009, bed night fees could no longer be paid directly to Lolkisale village and were instead paid to the Wildlife Division. Sulle and Banka (2017) documented that the revenue Lolkisale received dropped off a cliff from $95,003 per year in 2009 to $5,304 in 2010. This finding seems to run contrary to the notion that WMAs reflect a form of "community-based" conservation. However, it is also important to point out that Lolkisale had been undergoing a process of subdivision in 2010, so Lolkisale's former sub-villages began taking their own shares of the revenue, skewing Sulle and Banka's (2017) results somewhat. At the same time, data on Treetop's bed night payments between 2004 and 2014 reveal that the total revenue generated from bed night fees had actually been increasing over this period, but due to the new processes for central collection and taxation, the total payments received by the WMA declined after Randilen was formalized in 2012 (NTRI 2017). From the perspective of Lolkisale village, formalization of the WMA directly reduced their stake in Treetop's business by splintering their consistent revenue stream into a fragment of its once sizeable whole. After TAWA (30 percent) and Monduli District Council (5 percent) took their share, the WMA took half of what remained. The other half was then divided *equally* among the member villages. For the first few years after the WMA was established, from 2012 to 2015, the community payments were divided among Lolkisale, Nafco, Lemooti, Naitolia, Mswakini Juu, and Mswakini Chini. From 2015 onward, Lengoolwa and Oldonyo were included in the WMA, and the community income was split eight ways. As Sulle and Banka (2017) noted, villagers in Lolkisale became increasingly aware of these drops in revenue after formalization because the village council had to ask for greater contributions from people to pay for the construction of schools and other community development projects that were previously fully subsidized by bed night income. During my interviews in Oldonyo in late 2019, one Kisongo man (about fifty-five years old) voiced similar concerns. As he asked, "What is wrong with our WMA (*kampuni*) now? Can you tell me? Because in the past we never had to worry about paying for these kind of development projects, but now our council is always raising the issue of contributions to pay for things. I have no money to pay for these things myself, so this is why I am asking you, what is wrong with our WMA?" As this man's questions reveal, the history of high returns from Lolkisale's partnership with Treetops was leading to expectations that were no longer in line with the practical realities of the WMA's model of revenue distribution. The "milk of elephants" had begun to run dry, and Sulle and Banka (2017) expressed concerns that these changes might lead to declines in community support for conservation in Lolkisale. The issue of decreasing benefits relative to costs was exacerbated by the fact that Lolkisale had given up a sizeable portion of its land via Lemooti and was thus disproportionately affected by the costs of the WMA as compared to villages on the Naitolia side. This seemingly inequitable distribution of costs and benefits to WMA member villages is well documented in the literature and poses a particular governance challenge that is difficult to address through the current legal framework (Homewood et al. 2015; Sulle et al. 2011; Sulle and Banka 2017).

At the same time, in the pursuit of establishing community-based conservation areas across connected landscapes, the benefits of equally distributing tourism revenues across numerous villages allows those without suitable areas for lodges or camps to also tap into capital flows. This is especially important for villages like Nafco, where elephant dispersals bring sizeable costs to local cultivators, though the landscape is not as aesthetically appealing to investors. As I have observed in Randilen, despite some initial growing pains in Lolkisale, the community finds the benefit-sharing scheme across villages in line with pastoral forms of reciprocity. Unsurprisingly, herders need little convincing of the importance of sharing access to unevenly distributed resources. One Arusha man (about forty-four years old) from Mswakini Juu articulately described this dynamic during an interview in 2019: "For pastoralism and wildlife conservation, you must recognize the value of the ecosystem as a whole. You cannot divide things into specific pieces because you may find that this area has good grass, but another area has a river. So, it is hard to say that a village that contributes more land to the WMA should be compensated more because it depends on the distribution of significant ecological features. We have to work together to manage the area as a whole." The man's explanation highlights the importance of thinking about WMA benefit-sharing in the same way that herders think about semi-arid rangelands: reciprocity and collective institutions serve the economic interests of pastoralists in ways that fragmented property relations cannot.

Ethnographic attention to the lived experiences of Randilen's residents makes clear that the WMA provides community members with a range of collective benefits including employment opportunities, revenue for local infrastructure development projects, increased access to healthcare and education, and support for preparing and attending funerals. These social benefits of the WMA surfaced organically during interviews with community members. As one Kisongo man from Lemooti (about fifty-three years old) described it, "The WMA is a good thing and we benefit from it a lot. We are given a certain percentage of money from the WMA that we use to build schools. Our dream is to build a hospital here with the money from the WMA. Although the money that we currently receive is modest, we use it for construction projects in our village and it helps us a lot."

Interviewees consistently alluded to the equitable distribution of WMA benefits as a major factor underlying their support for the WMA. As one Arusha woman (about sixty years old) from Mswakini Juu said during an interview, "The WMA is a great resource for us because it helps us and moves together with us. It helps our community by sponsoring our children, especially those with disabilities who are not able to go to school because of poverty. The WMA helps those families. When we want to build a classroom or a school, the WMA helps us fund the construction." Another Arusha man from Naitolia (about fifty-four years old) expressed similar sentiments: "The WMA helps us a lot. They fund scholarships to sponsor our children and they help us build schools and purchase furniture for classrooms. These developments would not have been possible without the money that we get from the WMA, so I like conservation because of this." Community members spoke highly of the range of social services provided by the WMA that ease the challenges of everyday rural life. These social benefits across villages are linked to Randilen's business model and the important role that investors play in generating revenue from tourism.

The emergence of other tented camps and lodges in Randilen, besides Treetops and Boundary Hill Lodge, add further economic complexity to the scenario. Prior to the WMA, five tourist accommodation areas were demarcated in the Tarangire Conservation Area management plan. Three of them were permanent facilities (Treetops, Boundary Hill Lodge, and Naitolia Camp), one was a year-round camp (Sidai Camp), and one was a special campsite (Sand River Campsite) (King 2009). In total, those areas could host up to ninety guests. When the WMA was formalized in 2012, the tourist facilities changed shape. Naitolia Camp and Sidai Camp were discontinued, though Treetops and Boundary Hill Lodge endured. Naitolia Camp was later redesignated as a public campsite called *iltepes*, the Maa name for a common species of Acacia tree. A few other lodges and tented camps have been added recently, including Nimali Lodge at the former Sand River campsite in Lemooti, Kirurumu Tarangire Lodge at the former Tamarind Campsite in Mswakini Chini, and Ecoscience in Mswakini Juu. These lodges are considered part of Randilen WMA, and each offers a slightly different product. Nimali provides a relaxing experience for tourists, as it is nestled into the side of the river, and guests are lulled to sleep by the calming sounds of the babbling water at night. Kirurumu's ten tented cottages are set in a wetland area adjacent to Tarangire NP, in Mswakini Juu, with numerous buffalo around. Kirurumu, however, is challenging to access from the main parts of the WMA due to poor road infrastructure. Ecoscience comprises a tented conference center and luxury lodge established by a volcanologist who studies the geology of the rift valley volcanoes.

In 2023, the drawn-out conflict with Boundary Hill Lodge finally came to a close. Though the Australian investor passed away in 2018, the case had remained unresolved in the courts through 2022. Alleging that the investor's brother had been bringing guests to the lodge under the table, Randilen staff sent armed VGS to Boundary Hill to barricade the lodge and force him to leave. In 2023, Randilen attracted a Ugandan investor to take over administration of the lodge, who hired a lawyer to assist with carrying the case through. The case was finally closed in 2023 with the courts siding with the community. Randilen's management team then wrote a letter and delivered it to the investor's brother, stipulating that the land formally belonged to the WMA and that he would not be allowed to return. As of 2024, the Ugandan investor plans to run the lodge under the name Mwamba Lodge.

Other than the dispute with Boundary Hill Lodge, the community appreciates Randilen's lodges for generating revenue for their WMA, and there are no other outstanding conflicts that I am aware of. Local Arusha in Mswakini Chini, for instance, are grateful for Ecoscience's well-intentioned gesture to fund the construction of a dispensary and dining hall for its primary school. Community members I interviewed spoke of the value of having quality lodges in their WMA and often encouraged me to help search for more investors and advertise the area to prospective tourists. As one Arusha man (about thirty years old) from Mswakini Chini expressed during an interview, "The WMA has many benefits. The village office was one of them because we built it with WMA funds. We get money because of those investors inside the WMA who support us. The WMA staff wait at the main entrance gates and collect the money from the tourists staying in the WMA, and then they share it with the villages." Other interviewees expressed similar views. In the words of another Arusha man (about fifty-four years old) from Mswakini Chini:

> The WMA gives a lot of help to our community like building schools and dispensaries, but it depends on money coming in from tourists. We need more help from investors to strengthen our WMA and make sure it keeps benefiting our community in the future. Sometimes I worry that if there is no money coming in and the WMA does not receive enough support, then it might end. This would be a terrible loss for our community. I cannot predict the future, but I hope it will continue to bring in tourists so that our community can continue to benefit. Right now, there are NGOs helping to pay the VGS staff, so we need more support in the future so that Randilen can stand on its own.

As this narrative segment highlights, community members recognize the centrality of tourism to the WMA's business model. Given its positive impacts on their lives, Randilen's residents are hopeful that the WMA will increase its revenue so that it can continue to deliver meaningful returns to the community in the future. As one Kisongo woman (abour forty-two years old) from Oldonyo also expressed:

> I love the WMA because we are benefiting a lot from it. You know, our village was new, so we had no village office. We just held community meetings under the tree. That was our office. Now we have the village office. Loosikitok sub-village has built a primary school there with the money from the WMA. Lengijape sub-village also has a primary school built with the money from the WMA and now we are in the process of constructing a health

dispensary over there too. Those developments were made possible by the money from the WMA. So, I love the WMA because of these benefits that our community gets.

Residents of Randilen are happy that WMA revenue can be put toward impactful community initiatives, particularly if their village was not receiving bed night payments prior to the establishment of the WMA. Interviewees also consistently conveyed appreciation for the added employment opportunities that the WMA generates for its local community. As one Arusha woman (about thirty-seven years old) from Mswakini quipped, "Of course, I like the WMA because my husband works there as a VGS! *laughs* so I cannot say that the WMA does not have benefits because the income that our household gets is from the WMA." Similarly, as described by a Kisongo woman (about sixty years old) from Lemooti, "There is a big benefit to the WMA. My son works for the WMA as a VGS, so our money comes from Randilen. When the WMA first started, I was unsure how it would affect us. But now, I appreciate it because of the employment opportunities that it generates for our community."

One concern that was raised by the WMA chair in 2019 and several other interviewees, however, was that community members are not fully represented through employment at the lodges and camps. As per the revised management plan, community members are entitled to "priority on any employment opportunity" within the WMA, "provided they hold the required qualifications" (RWMA 2018, 16). My observations and informal conversations with the staff at the lodges and camps made clear that the *askari* (security guards) are generally local Kisongo men hired from Lolkisale-side villages. Kitchen and service staff, however, are almost always from outside the community. When I raised this issue with the lodge manager of Boundary Hill Lodge during an interview, he made a valid counterpoint: the luxury lodges must cater to wealthy tourists who expect five-star service. This makes it difficult to hire community members as chefs and servers in these lodges without adequate professional training. On the whole, residents of member villages end up taking up positions as security guards. Most employment opportunities generated by Randilen take the form of VGS positions with salaries paid directly by the WMA. The skills gap limiting local employment in technical positions is an issue in need of further attention. Increased education and capacity-building initiatives to prepare community members for diverse employment opportunities in the luxury lodges would serve to improve local representation in Randilen and the tourism sector more broadly.

Randilen's tourist lodges make varied economic contributions to the WMA. During interviews with the financial staff of Randilen WMA in 2020, they explained that Treetops still does most of the heavy lifting in terms of tourism revenue for the WMA, though Nimali and Kirurumu also make decent contributions. Ecoscience, however, is struggling to bring in tourists and researchers, and one of my interlocutors suggested to me in 2020 that the owner has been looking to sell the lodge. Investors also face the challenge of a never-absent central state that can decide to "shake the tree" at any given time. During a visit to Randilen's lodges and camps in mid-2020 to carry out interviews with the managers, a TRA vehicle was just ahead of me on the road visiting each facility in the same order. When I asked the manager at Ecoscience what the TRA representatives wanted, he replied jokingly that "sometimes they want to tell you your furniture is out of place, so you must pay a fine," implying with some sarcasm that TRA frequently hassled the lodges with unnecessary fines even if bookkeeping was in order. The macropolitical environment of private business in Tanzania is thus highly challenging because of the strong presence of the state in all aspects of private enterprises. This dynamic is rooted in the legacy of centralized resource governance policies that became entrenched through Tanzania's colonial and socialist periods (chapter 1). Despite neoliberal economic reforms in the 1980s, the state has consistently demonstrated an unwillingness to give private investors free reign—their primary purpose from a statist perspective is to generate revenue for central coffers through productive use of land and resources. This political issue exists in Tanzania irrespective of WMA status.

From the perspectives of lodge managers, garnered during interviews in 2020, the WMA structure has benefits and drawbacks. Changing bed night fees can detrimentally affect the bottom line of lodges. If an investor, for instance, had a prearranged agreement to pay a village a bed night fee of $5 prior to a WMA being established, and was subsequently contracted to pay a $15 bed night fee to the AA after the WMA was formalized, the investor has an incentive to oppose the WMA structure. Another drawback is that the added permit fees can deter some tourists who are looking to reduce trip costs wherever possible. Based on my own informal conversations with tourists, some seem content to pay a little more if they are assured that their money will go toward supporting local communities and the prospect of conservation outside national parks. At the same time, tourists can also be deterred by TAWA's bureaucratic process for acquiring WMA permits, which from the perspective of investors and tourists adds unnecessary red tape. Though it is under revision at the time of writing, the current system is rather convoluted and creates a significant barrier to entry. To enter Randilen WMA, a prospective tourist can either have their tour operator or lodge arrange the permits on their behalf. More adventurous tourists can request a permit on Randilen's online website by filling out their vehicle registration number, names of visitors, and passport details. Randilen's administrative staff then produce an invoice that includes a control number that will be linked to a subsequent payment. The permit fees can then be paid either with an MPesa mobile money transaction or at an NMB bank in Arusha town. Once the WMA is provided with the proof of payment, Randilen administrative staff will email the tourist the permit. This arduous process, inhibited by an overly bureaucratic central state, creates unnecessary hindrances for tourists who might otherwise want to simply purchase their permit at the entrance gate. Despite rumblings of reform, purchasing WMA permits at the entrance gate was still impossible at the time of my fieldwork. Communication between entrance gates and lodges is also sometimes strained by poor mobile networks and a lack of coordination. In some cases, this results in tourists getting stuck at the gate without a permit in hand because the lodge was expecting them, but the gate staff were not made aware. The lodges are also challenging to operate due to their remoteness, which would be an issue irrespective of the WMA. Poor road infrastructure and black cotton soils currently make wet season travel into the heart of the WMA rather daring.

At the same time, lodge and camp managers also noted positive aspects of the WMA structure. The WMA provides lodges and camps with greater security, as some of them were plagued by petty thefts prior to formalization. Randilen has two well-staffed ranger posts, which serve as deterrents for prospective thieves. Wildlife populations have also been on the rise since establishment of the WMA, enriching the product that lodges and camps can offer to clients (Lee and Bond 2018). This has been further enhanced by the development of tourism infrastructure, like the marvelous elephant hide near Sunset Hill, and the newly built airstrip for bush planes. These facilities could serve to attract more tourists in the future, so long as the WMA is able to sustain itself economically.

Economic Viability of the WMA Model

The formal benefit sharing model of WMAs codified by the state does not fully account for the administrative costs necessary for operating WMAs. When Honeyguide commenced work in Randilen, it brought with it a clear focus on making the WMA economically viable, a pursuit that has been constrained by legislative context. Having worked as a photographic tour operator in Ololosokwan in the 1990s, Damian was very familiar with the economics of community-based conservation prior to the legal reforms of 1998. He was determined to help Randilen develop into a community-owned and operated business that was financially independent, and which provided opportunities for community livelihoods, wildlife conservation, and tourism to thrive. This meant ensuring that Randilen's gross tourism revenues covered its operational costs on a consistent basis without external donor support. One of the most challenging issues in realizing these objectives, as voiced to me by Randilen's finance team during interviews in 2019–2020, is that TAWA collects 30 percent of the WMA's gross revenues without contributing anything back on the ground in terms of service. In 2016, new nonconsumptive tourism regulations were introduced for WMAs, which reduced centrally collected revenues marginally from 30 percent to 25 percent for TAWA (while maintaining the 5 percent for district government). Accordingly, Randilen WMA should keep 70 percent of its own revenues, though it was only receiving 65 percent at the time of fieldwork. Put differently, TAWA takes unsustainably without helping to defray the management and governance costs of operating the WMA. My recommendation here to government policymakers is a reduction in the percentage of revenues kept by TAWA as compared to the WMA. This will likely help generate *more* money for the government over the long term by helping foster economically sustainable WMAs.

The current model makes Honeyguide's goal of helping the WMA achieve financial independence challenging, but not unattainable. The money designated for WMA operational costs must cover governance, management, and protection practices, which is no small task (NTRI 2017). As a financial breakdown, in 2014–2015, Randilen WMA earned $163,387.69 in total gross revenues (NTRI 2017). From this total, $49,016.31 was first taken by TAWA, and $8,169.38 was taken by Monduli District Council. From the remaining $106,202, half went to the WMA to cover Randilen's operational costs ($53,110), and the other half was divided between the member villages, which amounted to roughly $6,640 per village per year (NTRI 2017, 2). Based on a financial assessment of the WMA by Acacia Natural Resource Consultants in 2016, it was determined that about $176,000 was needed to operate the WMA. Honeyguide developed their own "WMA Financial Viability Tool" based on this initial dataset. Damian has since suggested to me that with inflation, Randilen would need to generate around $350,000 annually to achieve financial independence. Thus, Randilen WMA was only able to cover about 30 percent of its own expenses, leaving a deficit of around $124,198 between its earnings and running costs (NTRI 2017). As a rough breakdown of Randilen's operational costs, financial and administrative management expenses take up about 20 percent of its budget. Protection costs, including VGS salaries, patrol vehicles, fuel, training, and equipment run about 60 percent of its budget. Governance expenses, including venue rentals for AA meetings, transportation for association members, and seating allowance per diems make up about 20 percent. It is easy to overlook the fact that with all of these running costs, money coming into the WMA must be enough to cover money being spent, and this makes it a challenging business to operate, even for a professional financial management team. Due to Randilen's financial deficits, its governance and management activities must be supported by Honeyguide with money from the Nature Conservancy and international donors to help it fulfill its operational tasks. Thus, as things stand at the time of writing, the viability of the WMA is still subsidized by philanthropy.

This consideration highlights the fact that the current level of government taxation is jeopardizing the long-term viability of Randilen WMA. As one member of Randilen's financial team told me during an interview, "The central government does not realize that Randilen WMA is a fruit-bearing tree. Rather than eating the fruit over a long period of time, they want to cut its branches for a quick profit." This sentiment was vocalized by WMA staff and community members, some of whom viewed the central government as nearsighted in its attempts to maximize short-term profits at the expense of long-term sustainability. Put in more diplomatic terms, the central government likely needs to take a little less and support WMAs a little more if they are to become a successful model of conservation across the board. It is worth reiterating here that Randilen is a young WMA, and its year-over-year revenue growth is impressive and trending toward financial sustainability.

A lack of harmonization across the arms of the Ministry of Natural Resources and Tourism (MNRT) also bears on Randilen WMA's viability. The Tanzania National Parks Authority (TANAPA), for instance, did not allow multiple-entry visitors into Tarangire National Park at the time of my doctoral fieldwork, as it wanted to encourage tourists to stay inside the park. Single-entry permits make it highly inconvenient for guests staying at lodges in Randilen to visit Tarangire NP as part of their safaris because each time they leave the park to return to their lodge, they have to pay for a new twenty-four-hour single entry permit. Making matters worse, TANAPA refused to build a ticketing booth at the Boundary Hill Gate connecting Randilen WMA to Tarangire NP. Tourists who would have otherwise been able to drive for a few minutes from Boundary Hill directly into the park must instead drive all the way out of the WMA, along the main highway, and back through the main gate of Tarangire NP to purchase their single-entry ticket. They can then drive back to their lodge in Randilen via the Boundary Hill Gate, which does not restrict movement out, but will not allow tourists to enter without purchasing a ticket at the main gate. Recently, tourists in Randilen WMA have been able to drive from Boundary Hill back to the airstrip at Sunset Hill to purchase a TANAPA ticket, making for a slightly shorter two-hour-drive detour. The commitment it would take for TANAPA to implement a ticketing booth at the Boundary Hill Gate is fairly modest compared to the massive inconvenience it causes for prospective tourists, who are largely drawn to Randilen WMA because of its proximity to Tarangire NP. As described to me by several of my key informants, however, TANAPA's reluctance to implement the booth is likely because it views Randilen as a competitor, and it wants to disincentivize tourists from staying there. Tourism revenue generated from visits to WMAs, after all, is collected by TAWA and thus falls outside TANAPA's reach. TAWA's overly bureaucratic system for administering WMA permits compounds the potential inconveniences experienced by tourists. Closer collaboration between TANAPA and TAWA would improve the Randilen WMA visitor experience and help streamline Randilen's product on offer. Harmonizing the two arms of the ministry is becoming especially important since Tarangire NP is now saturated with lodges and camps and has no more room for investments inside the park. Thus, demand to invest in the adjacent WMA will likely continue to rise in the future.

International NGOs and donors are aware of the constraints posed by macropolitical context. The shortcomings of the central government in supporting WMAs create opportunistic gaps that must be filled by strategically positioned NGOs, meaning that the status quo finds a way to sustain itself in a way that benefits the government and the NGOs who pick up the slack. But the question that Honeyguide is determined to answer is, Can a WMA achieve true sustainability on behalf of its community? The Northern Tanzania Rangelands Initiative (2017) suggests a few ways in which the government could help. First, TAWA must consider reducing its revenue collection from WMAs. For WMAs to become sustainable, it will likely be necessary that they simply keep more of the tourism revenue that they generate. NTRI (2017) also recommends streamlining the framework for investing in WMAs to establish more lodges and tented camps, and improve roads and accessibility. Of course, they also advocate for the continuation of outside donor support to supplement tourism revenues since WMAs are currently unviable without this support. Almost all WMAs in Tanzania rely on donor funds to pay VGS salaries and purchase patrol equipment (Kimario et al. 2020).

Under the guidance of Honeyguide, Randilen had been working admirably toward the goal of financial independence when the COVID-19 pandemic hit in March 2020 and temporarily derailed its growing revenue stream (Shoo et al. 2021). Fortunately, Randilen's revenues have since recovered. Thinking critically, there are some significant issues that make it challenging for Randilen to reach complete sustainability. Treetops lodge currently generates the majority of the WMA's revenue, though it can only host a maximum of forty people per night. This means that in an entirely hypothetical best case scenario, with a "full house" every night of the year, and considering the $15 bed night fee, the maximum revenue it could generate through bed night payments is $219,000 (40 people × $15 a head × 365 days a year). Permit fees generated from these hypothetical visitors (40 people × $10 × 365 days) would total $146,000. In reality, of course, a full house every night is impossible given seasonal closures. This maximal figure ($365,000) would then be reduced by 35 percent ($127,750) by the government, and half of the remainder ($237,250 / 2 = $118,625) would be distributed to member villages. Assuming an overly modest estimate of $200,000 a year to cover Randilen's running costs, Treetops alone simply cannot carry the entire WMA on its shoulders. For Randilen to become economically viable without donor support, it needs more investors.

Importantly, it is not just the *number* of investors that matters. The *quality* of the investors, measured in terms of Randilen's economic and environmental sustainability goals, is also of great significance. In Burunge WMA, for instance, one lodge in Mwada village sleeps sixty-five tourists per night and generated approximately $268,9000 per year in 2016, while holding a small concession, compared to another lodge in Vilima Vitatu ("three hills") village with a large concession that only generated about $50,400 for the WMA per year (Moyo et al. 2016). The point, then, is that it is not just about increasing the total number of lodges, but growing the number of tourist accommodations that bring in significant bed night and permit fees without unsustainably using available land. Spearheaded by Honeyguide, the Nature Conservancy, and the Land & Life Foundation, tourism occupancy rates in Randilen WMA increased by 48 percent between 2017 to 2019. At the time of my doctoral fieldwork, Randilen had two sites open for potential future investors (Sunset site and korongo la Dume site) that could each sleep an additional twenty to forty guests per night. As of follow-up fieldwork in 2024, Randilen had confirmed two new investments from Sunset Tarangire to build a luxury five-star hotel, and Entara to establish a tourist lodge in these sites. The AA has since identified three more sites for potential investors—one near Kirurumu and two others close to Boundary Hill. Prospective investors who wish to establish a lodge or tented camp in these sites must negotiate with the WMA's bid and tender committee comprising the AA, Randilen's management team, and representatives from central and district government. The contract would include an initial and recurring fee to the WMA to operate within its boundaries, which could be a drag on the investor's margins but a boon for the WMA's revenue.

For Randilen to become financially independent, a few more quality lodges or tented camps would help a great deal. Building a consistent revenue stream for the WMA that exceeds its basic operational costs would create opportunities for reinvesting capital into aspects of the WMA that would improve its overall efficiency and in turn attract more tourists and operators until saturation is reached. The revised management plan of 2018 included an ambitious goal to generate $2.5 million annually by 2020, which Randilen was admirably trending toward prior to being disrupted by the global pandemic (RWMA 2018). Since the return of tourism, Randilen has been growing its revenue year over year, but still has a little way to go to reach its ultimate goal of sustainability.

Corruption and the "New Nafco"

Another important consideration in the current benefit-sharing structure is the role of local governance at the village level in ensuring that income from the WMA is distributed equitably to the community. Local governance played a significant role in the era of private contracts between villages and operators prior to the WMA. Since the centralized reforms, it has remained crucial for ensuring that the tourism income that is returned to villages is managed transparently and in a way that is reflective of assembly interests. This is particularly critical if there are limited funds available for each member village, which necessitates careful budgeting and accountability practices.

As discussed earlier, Lolkisale village has been fortunate to have good governance at the local level since its establishment, and this has meant that wildlife-related income from its Treetops partnership has funded meaningful community development projects at the village level. Even after its income declined drastically following the legislative reforms, the village council continued to act democratically in allocating these funds for community initiatives. Following in the footsteps of Lolkisale, the leadership of Randilen's most recent village, Oldonyo, has fostered local support for Randilen through strong and transparent governance institutions. As one Kisongo woman (about fifty-three years old) from Oldonyo voiced during an interview, "The key to the WMA is our leaders. They are the ones who give us information about the WMA from the meetings. We love the WMA because it helps us build schools through contributions to the village. We trust the WMA because our leaders bring to our assembly all matters and issues, so we always know what is happening."

While Lolkisale and Oldonyo villages illustrate the positive impacts that transparent governance at the local level can have on the relationships between WMAs and villages, the experience of Nafco village paints a very different picture. Having subdivided from Lolkisale in 2012, Nafco comprised a rapidly growing Arusha majority that had been expanding land under cultivation, primarily for maize, beans, and peas. Unlike Lolkisale, Nafco did not have a history of managing money from wildlife-related tourism, and the village council that immediately came into power after subdivision was tasked with managing revenues after they were distributed from the WMA to the village bank. Although the annual funds were modest (about $5,000), the community was aware that this money had been allocated to the village. At first, the assembly trusted their elected leaders and assumed that the money would be used to benefit the community through the construction of schools and dispensaries, which were high on the list of infrastructure priorities that the community put forward in assembly meetings. The Nafco council explained that the money they had received was not enough to fully fund those types of projects, so village residents would have to continually supplement the money with their own contributions. People obliged to this form of taxation, as they were confident that it would improve quality of life in their community. As time passed, however, the village council continually asked for contributions on top of the WMA money, and Nafco residents did not see returns on their investments in the form of community development projects. People began to grow resentful and felt that the village council may have been embezzling money. As one angry Arusha man (about forty years old) explained to me in Nafco in 2020, "The village chair asked us all for our contributions with a smile on his face and promised that the money would be used together with the WMA funds to build a school, but then he embezzled the money!" As described by another interviewee, the village chair had been seen driving around in a new 4×4 vehicle that he had allegedly purchased with villager contributions and WMA income. Unlike the ambulance in Lolkisale that the community had voted upon as an appropriate use of funds, the Nafco leader had decided undemocratically to use the money to buy the vehicle. Furthermore, Nafco residents felt that he used the vehicle for his own private needs and not as a public service to the community. One Arusha woman (about sixty-two years old) was deeply upset in 2020 when asked to reflect on the past leadership group:

> I will not speak a lie. It is better to speak the reality. Our previous village leadership was not good. They embezzled all the money that the WMA was giving to us. If you see the health dispensary and the village office, we are the ones who funded their construction by giving our own contributions. We asked our leaders, where is the money from the WMA going? Now we have new leaders and we are waiting to see if these new leaders will show people that the WMA is giving the money back to the community. We are keeping our eye on them to see how they will govern our village. We need to see more development from that money. When I think about the past leaders, I feel like I want to cry because it was not just the money from the WMA, but our own contributions that we gave them willingly, and they robbed us. I stopped myself from buying clothes so I could give my contributions in support of the village development work, but that money was embezzled. They took our money and never gave us anything back in return. They were not transparent leaders. They were hiding their sinful deeds.

As I continued to garner people's perspectives on local governance in Nafco, one young Arusha woman reasoned that she was not sure whether the chair had explicitly embezzled the money or whether he had simply mismanaged it. The consensus that I seemed to gather was that the village council had acted in a matter that was secretive and nontransparent, and with little accountability to the assembly that it represented. This caused significant distrust to build up in Nafco, not just toward the village council but also toward the WMA, which people began to see as an institution that benefited those in power and not the general village assembly. The fact that the village chair had asked for continued contribution payments from Nafco residents made people's experiences of resentment so much deeper. Negativity was still apparent in the survey responses I gathered in Nafco in 2020: 10 percent ($n=11$) disliked the WMA, and 14 percent ($n=16$) were neutral—proportions that were notably higher than in other villages.

Many of the people I spoke to in Nafco about the corruption were not just bitter but hurt. They had trusted their leader to act in their community's interests, and he had abused his position of power for his own benefit. By betraying people's trust in this fashion, the village council was not just undermining local governance, but the effectiveness of the WMA, as Nafco was prevented from reaping its economic benefits. When community members saw the village chair driving around in his new car, it was the straw that broke the camel's back. The assembly rose up in revolt and stormed the offices of the council and chased them out of town. Not a single member of any of the village's committees was spared. When I carried out fieldwork in Nafco in late 2019, prior to the elections, Nafco had no council or chair. The only member of government there was the *mtandaji* (Village Executive Officer) who was assigned to the village from the district to carry out administrative tasks on its behalf. She informed me that just prior to my arrival, officials from the Prevention and Combating of Corruption Bureau, a national body for investigating government corruption, had visited her to gather information about the case and the potential whereabouts of the former leaders. Knowing that elections were a particularly significant topic in Nafco, I stayed in the village while elections were ongoing in November 2019. During this period, there were massive crowds and much political fanfare, as Nafco residents made it very clear that they would not tolerate corrupt governance in the future.

The contrasting cases of Lolkisale-proper and Nafco demonstrate the importance of village governance in shaping the social outcomes of WMAs. In Lolkisale, transparent and democratic use of funds led the assembly to trust its leaders and realize community-level benefits from the WMA. By contrast, Nafco leaders managed WMA funds secretively and used them undemocratically for personal gain. Consequently, the community developed distrust for their leaders and for the WMA institution, which they came to view as a tool for further enriching those in power at the expense of those who were already impoverished. Ironically, the first chair of the WMA elected by the AA was from Nafco village during this period. He was deeply disturbed by the mismanagement of funds at the village level. Despite his best efforts to improve flows of capital from the WMA to the village assembly by pressuring Nafco's leaders to change their ways, he was ultimately unable to make headway and decided to resign. In his words,

> I was the first chairman of Randilen WMA chosen by the AA. But I was not happy being the chairman and seeing how my village was misusing the money they were receiving. I tried to advise the village leaders to build a village office. The leaders told me no, we have to buy a car. So what's more important, the office or the car? The office is more important because you keep all village records and everything in the office! So, I had a conflict with those leaders. I was not happy being the WMA chair and knowing that the money was not benefiting my village, so I wrote a letter and resigned.

The chair's decision to resign from his post demonstrates that WMA institutions are still contingent upon village-level ones, which determine whether tourism revenue from WMAs is equitably distributed to village assemblies. Village councils are gatekeepers with the power to shape local sentiment toward WMAs by either facilitating flows of capital to community development projects or mismanaging money for personal gain. The former approach builds local support for conservation, while the latter leads to disenchantment.

The repercussions of poor village leadership were still observable in Nafco in 2019–2020, as manifest in people's attitudes toward Randilen, though people had softened to the WMA considerably thanks to Honeyguide's efforts to help reduce the impacts of elephants on crop production. These initiatives have helped to rebuild community trust in the WMA and remedy the damage done by the former leaders. Notably, most of my survey respondents (63%; $n=72$) in Nafco reported in 2020 that they trusted Randilen's authorities to act in their community's interests. While this proportion was lower than in other villages, it clearly shows a rise in community trust and suggests that the relationship between the WMA and Nafco is in repair. Some interviewees even expressed great appreciation for the WMA. One Arusha woman (about forty-eight years old) in Nafco said, "We love Randilen WMA. When I say that I love the WMA, this is because when there are any problems like a funeral, they are the first ones to come to the village. And we are hearing from the meetings that there is money that we are getting from the WMA and that money is now being properly used for our benefit. So, to be honest, I can't say that I dislike it when I see some of the benefits that I get." During interviews with the new village chair of Nafco in 2020, he pledged to help cultivate "a new Nafco" that respects the values of democracy and community well-being. As voiced by other members of the newly elected council during interviews, Nafco's political leaders have come to see the WMA as an institution that can support their community, as it had Lolkisale. One of the new sub-village chairs had this to say:

> The problem in this village is that the previous village government was not good, so the money from the WMA was not put to work for the community's benefit. We saw that our neighbours, like Lolkisale and the other villages, had made much effort with their development activities with the same amount of money. Those villages were doing great, but for us we did not see the fruits of that money because of poor village leadership. We are trying our level best to restore people's trust and show that this new leadership group will use the money from the WMA transparently for the community's benefit.

Similarly, as echoed by another sub-village chair, "The previous village leaders were not transparent to the assembly. There was no cooperation with community views during village meetings. They just took decisions based on their own perspectives. After the new elections, we agreed that we have to listen to the assembly and try to build trust by involving people in decision-making processes about village development." A consistent theme in these interviews was recognition of a moral obligation to do right by their assembly and "succeed where the past leaders had failed." The survey and interview data presented here show that Nafco has been trending in the right direction, but time will tell how the new leadership group performs in the future.

Socioeconomic Effects

Another layer of complexity within the WMA model is the question of how conservation practice impacts household economy. Existing economic studies have made the case that WMAs generally do not increase the income of individual households (Keane et al. 2020) but that they do in some cases provide community-level benefits in the forms of improved infrastructure and the construction of health dispensaries and schools (Homewood et al. 2020). My ethnographic research on Randilen WMA complicates this literature.

One apparent aspect of WMAs that quantitative economic studies seem to overlook is the manner in which individual household economy and community-level infrastructure development are intertwined. As I observed in Oldonyo village, if tourism income does not cover the costs of community development projects, then village leaders will go door-to-door to the assembly to request payments to cover these costs. Thus, while WMAs may not *directly* improve individual household wealth, they may do so indirectly by covering costs that villagers would otherwise have to pay to their local governments in the form of "contributions" (*michango*). Community-level and household-level economic benefits in such instances become intimately linked. Support for children to attend school in the forms of scholarships, classroom construction, or provision of school supplies also do not directly create an increased flow of capital from the WMA to households, but they nonetheless indirectly reduce economic burdens on households to cover education costs themselves. An Arusha man (about sixty-one years old) from Lengoolwa spoke to this point during an interview: "We can see lots of ways that the WMA is helping us. Our village office was built from funds from the WMA. If it had not been for the WMA, we would have still been making contributions to build the village office. But because of the WMA, the village office is there." As the man describes, WMA revenue replaced the need for community-level taxes to build the village office, thus indirectly increasing household wealth.

Furthermore, it is also worth revisiting the conventional wisdom that conservation produces inevitable trade-offs between ecological outcomes and human well-being (McShane et al. 2011). When I interviewed community members about Randilen, there were only a few individuals who reported disliking the WMA. Those interviewees usually referenced fines for grazing livestock in the reserve area without permission. Interestingly, when I prodded more in those interviews, the individuals had not personally grazed their livestock in the WMA but had only heard accounts of fines from others. The secondhand stories were enough to color their attitudes in a negative way. Importantly, then, the few interviewees with negative opinions did not actually consider themselves to be WMA resource users. Of all the interviewees who grazed their livestock inside the WMA, every one of them liked it. In the words of one Arusha man from Naitolia (about thirty-four years old), "Conservation is good because if they had no conservation around this place, there would be no grass for livestock grazing. I am happy that the WMA is conserving the environment because now our livestock have lots of grass around." Narratives like this one, expressing gratitude for the grazing bank, complicate the idea that conservation restrictions and livelihood benefits are mutually exclusive (McShane et al. 2011). In this case, the resource users who are directly affected by conservation enforcement are the ones who appreciate the WMA because of its positive economic impacts on their livelihoods.

It is important to critically consider the measures that existing economic analyses of WMAs employ to quantify household-level benefits. Complex statistical models may ironically exclude the most important economic factors that are not easily quantified—the direct livelihood benefits of a well-managed pasture and crop-protection services. Randilen delivers significant returns to local livestock keepers through a locally governed and flexibly enforced management model that keeps as its primary focus the pastoral mode of production. One Arusha man (about forty-one years old) in Mswakini Juu explained this well: "If you ask me what the benefits are of the WMA, I can tell you for sure that I am not going there to take photographs of wildlife! I see enough wildlife near my boma at night! *laughs* But the real benefit that is difficult to put a price on is the livestock grazing area. That is the biggest benefit our community could ask for." What seems to be the most important factor underlying community support for Randilen WMA is the direct contributions it makes to local livelihoods through its communal grazing bank and crop protection services. Quantifying the economic impacts of Randilen WMA on household wealth, as Keane et al. (2020) attempt to do across numerous WMAs, would be an exceptionally challenging endeavor in my view. My suspicion is that the household wealth indicators they used would not fully represent Randilen's social-ecological complexity.

What matters most to local Kisongo in the context of Randilen WMA is the management of grass, which sustains their cattle. Grass is the material base of their economy, and cattle are the currency. If there is good grass, and livestock are healthy, then people consider themselves to be wealthy (*engishui sidai*—"a good life"; see Woodhouse and McCabe 2018, 5). For a rigorous analysis of the economic impacts of WMAs to hold water in a pastoral context, it would have to consider the precise economic benefits of a well-managed common pasture. The Arusha, who greatly value both livestock (to a lesser degree than the Kisongo) and farms, are much more inclined toward the material indicators of wealth that feature into classical economic surveys because they prefer private property and permanent settlements. The Kisongo, by contrast, still think of themselves as being semi-nomadic (despite sedentarization) and do not value modern houses in the same way. Thus, focusing too narrowly on traditional measures of household and material wealth might not tell the full story of how WMAs fit in relation to the economies of local communities. Certainly, there are numerous other relevant factors that would also be challenging to quantify, such as the impacts of increasing wildlife populations on livestock and crop production (Raycraft 2024a, 2024b). Perhaps ethnographic attention to the lived experiences of local communities should be considered on equal footing with meta-level quantitative analyses across cases.

The anthropological findings presented in this book suggest that from a community perspective, the value of Randilen WMA outweighs the drawbacks: 75 percent ($n=511$) of the people I surveyed about the WMA felt that Randilen had more economic benefits than costs. A minority (18%; $n=125$) reported that the WMA had an equal distribution of costs and benefits, and only a very small number (6%; $n=42$) maintained that it had more costs than benefits. This marker of conservation equity and justice is notable given the countless cases of top-down protected areas across sub-Saharan Africa.

Conclusion
Conservation at the Crossroads

In the context of Tanzania's conservation sector, the post-socialist state has attempted to configure its institutional machinery for resource governance to efficiently extract tourism revenue from wildlife resources. Against the backdrop of structural adjustments and public-private partnerships between foreign investors, nongovernmental organizations (NGOs), and the government, Wildlife Management Areas (WMAs) have emerged as a contemporary form of decentralized conservation governance and management. On the one hand, WMAs seem to allow the state to monetize wildlife that moves outside national parks, in effect recentralizing resource control by extending the reach of the state apparatus onto village land (Gardner 2016; Nelson et al. 2007; Wright 2019). At the same time, they also create new formal institutions for pastoralists to manage rangelands in the face of land alienation and for cultivators to articulate tenure claims.

In the case of Randilen WMA, the perspectives of local Arusha and Kisongo Maasai community members speak to participatory governance and management mechanisms, despite a complex history (chapter 2), ethnic frictions (chapter 4), and textured layering of land use institutions (chapter 3). These lived experiences have led to a rise in community support for conservation (chapter 5), seemingly reflecting a process of environmentality that has succeeded in bringing the interests of rural citizens in line with the state (Agrawal 2005). Careful examination of people's views, however, reveals that the central tenets of wildlife conservation do not resonate with community members, who are primarily interested in securing land. Thus, an environmentality framework, in and of itself, does not fully represent the resonating value of rural communities in maintaining their ongoing ways of life. More precisely, it seems that there is common ground between the interests of the central government, which views wildlife as a key source of revenue, conservationists who want to protect wildlife for intrinsic reasons, and rural communities who seek to secure access to land in the face of alienation and derive new livelihood sources.

I forward here again the notion of having *conservation in common* with reference to cases where a territory connotes different meanings to groups of actors with distinct aspirations yet simultaneously serves those stakeholders in valuable ways. Fundamental to finding commonality is empathy, an exercise in taking the perspective of others and seeing territory from a different point of view. Rather than a fortress model of conservation predicated upon forcing a vision of pristine nature onto a landscape at the expense of local and Indigenous communities, or an environmentality approach that attempts to forge people into subjects who think and feel the same way as those in power, I propose here empathetic conservation as an alternative conceptual framework. Empathetic conservation requires authorities to take stock of the things that matter to local communities and prioritize them as centrally important to the management model. As evidenced by the Randilen case, when community members feel that conservation protects the things that they care about, support for conservation emerges organically through feelings of reciprocity, trust, and mutual respect (chapter 6).

Making a concerted effort to find common ground can be an effective way to reconcile conflicts between conservation stakeholders. Cooperation is most likely when the objectives of different groups are not mutually exclusive, despite the diverse meanings they attach to conservation territories. In the case of Randilen WMA, community desire to secure pasture coexists with conservationist interest in protecting wildlife habitat. Both sets of actors benefit from the reduction of land fragmentation, which would threaten pastoral livelihoods and ecological connectivity. Increased capacity for protecting farms from encroaching wildlife, coupled with enhanced tenure security, has helped foster support for conservation from smallholder cultivators as well. Randilen WMA provides an example of different stakeholders having *conservation in common*. It does not, however, demonstrate that environmental subjects are produced by government, as Foucauldian logics might lead one to believe.

Despite the promise shown by Randilen, Tanzania's WMA model of community-based conservation is not exempt from scholarly critique. Following their introduction in the late 1990s, WMAs essentially functioned to foreclose direct investments on village land and recentralize revenue collection from wildlife-related tourism occurring outside national parks and game reserves. Strong evidence of mutually enhancing village-investor partnerships in Lolkisale and Loliondo prior to the introduction of WMA policy show that alternatives to Tanzania's centralized model of community-based conservation are possible (Gardner 2016; Nelson et al. 2010). The central government likely recognized that villages were becoming increasingly powerful in the context of neoliberal partnerships with private investors and viewed this development as a threat to state authority over valuable wildlife resources. In response, land laws that undercut state control were overruled by WMA regulations that weakened the sovereignty of villages to govern territory and negotiate with investors (chapter 1).

Despite the top-down way that WMA reforms were rolled out in Tanzania, what has become clear through this study and Wright's (2017, 2019) work is that Maasai communities have learned to wield them from below in defense of pastoral territory and in support of local livelihoods. Compared to other types of protected areas, WMAs are multi-use and feature significant opportunities for local communities to tailor them to fit their customary ways of life. This is especially significant for pastoral communities living in wildlife dispersal areas adjacent to national parks, in villages that overlap historical Game Controlled Areas (GCAs). In Loliondo, tensions over the ambiguous boundaries of villages and Loliondo GCA, and local resistance against the prospect of a WMA, have culminated in the state taking a stronghanded approach. In 2022, the central government demarcated Pololeti Game Reserve at the expense of local herders, who have now been dispossessed of key dry season pastures. Further yet, in 2023, the government proposed to formally upscale all GCAs in Tanzania to game reserves, a proposal that was currently being debated in parliament at the time of fieldwork in 2024. If put into practice, the policy change would displace dozens of pastoral villages in wildlife-rich areas of Maasailand. Fortress conservation in the form of centrally managed game reserves is thus a very real threat to pastoral communities in northern Tanzania, highlighting the significance of WMAs as perhaps the most viable option for securing tenure and continued access to rangelands in wildlife dispersal areas. Though some Maasai leaders and pastoral land rights NGOs are still pushing for the village institution and communal Certificates of Customary Rights of Occupancy to be recognized by the state, evidence from Loliondo suggests that the state is unwilling to fully devolve authority over lands it deems valuable for wildlife conservation to the level of villages. As shown by Enduimet in Longido and Randilen in Lolkisale, Maasai communities that have adopted WMAs on village land have been able to secure their livelihoods while satisfying government imperatives to form wildlife conservation areas.

In my view, this is clearly the lesser of two evils when faced with the prospects of evictions and dispossession.

Though WMAs have become codified as official conservation policy in Tanzania, economic and political complexities still affect their implementation in practice. Attempts to launch community-based initiatives founded on collective institutions face challenges in dealing with the role of domestic elites and wealthy individuals who stand to lose from increased environmental regulation (chapter 4). Capacity for communities to deal with rich herders, politicians, and commercial farmers is important. Furthermore, though WMAs replaced the direct investment model, village councils continue to play a crucial role in determining the flows of money from WMAs to village assemblies (chapter 7). Transparent and accountable governance practices foster support for conservation, while secretive and selfish leadership decisions engender discontent. Village institutions can thus undermine or reinforce the effectiveness and sustainability of WMAs.

Finally, the benefit sharing model of WMAs could use further attention from a policy perspective. Equal distributions of WMA revenues to each member village have trade-offs to consider since villages hosting investments receive the same shares as villages without tourism infrastructure. These dynamics may cause tensions in some contexts depending on local politics, though Randilen's community is content with the current arrangement. Reductions in the revenue percentage kept by the Tanzania Wildlife Management Authority (TAWA) would provide WMAs with more operational funds, helping them reach financial independence from international NGOs and donors. TAWA could further help by streamlining its process for tourists to purchase permits and collaborating closely with the Tanzania National Parks Authority to establish entrance gates between parks and WMAs as needed.

Perhaps most importantly, WMAs must deliver returns to the communities who invest their land and resources in them. This should entail the provision of societal benefits through improved social services, but also household-level ones through enhanced livelihood security. Demonstrating to communities that WMAs can sustain people's livelihoods is the key to unlocking local support for conservation. Ensuring that WMAs provide benefits that community members consider valuable requires attentive listening to people's concerns and a willingness to prioritize them in bylaw formation and conservation enforcement. Thus, their successful implementation may hinge on the extent to which they are designed and implemented in an empathetic way.

Cultural Politics of Transformation

I have endeavored in this book to situate Randilen WMA in relation to the intersecting social histories of Kisongo pastoralists and Arusha cultivators. These two Maa-speaking ethnic groups have closely related identities and complementary economies and have historically vied over access to and use of rangeland resources through warfare in the precolonial era and more recently through precise political negotiation of the administrative institutions of the post-socialist state (Kuney 1994). As Gardner's (2016) research demonstrates, ethnic territoriality has become more pronounced in Tanzania's Maasailand in the era of conservation neoliberalization. Randilen WMA has to some extent become an "arena for struggle" between pastoral Kisongo and agricultural Arusha (Neuman 1998). With the increasing encroachment of Arusha on the lowlands, the Kisongo view the WMA as a way to secure pastoral land from agricultural development. The Arusha were originally afraid that the WMA would displace their precarious "indigenous" claims to land, fears that were stoked by elites from outside the community via nefarious relations with the village councils. Once the dust settled after open protests, the Arusha eventually realized that the WMA in actuality reinforced their claim to land on the shoulders of their gained status as member villages of a conservation area. As agropastoralists who are integrated with the regional economy of the Kisongo, the Arusha are happy to keep livestock and reap the benefits of Randilen's grazing banks.

While both groups have, for the most part, come to support Randilen WMA for different overlapping reasons, there is still considerable apprehension about the future. Communities in northern Tanzania are always cautious about the potential for the state or private actors to prioritize wildlife conservation over local livelihoods owing to the lasting legacy of centralized resource control, and the history of dispossession that rural communities have endured. As voiced by a Kisongo woman (about fifty years old) from Lemooti village, "Randilen WMA is very important to us. That is our forest. It protects our livestock and wildlife together. We are happy with our WMA but we still worry sometimes about the future when we see what is happening with other protected areas in Tanzania. We rely on the WMA for livestock grazing so it would be devastating if the government took that area away from us." Some Maasai leaders are still skeptical of WMAs, viewing them as an extension of the state, and a gateway for the Ministry of Natural Resources and Tourism to later convert them to game reserves. These fears are grounded considering the history of fortress conservation in Tanzania, but my own conversations and interviews with government officials working in the wildlife sector point to little interest in upscaling WMAs en masse. As international pressure to consider the human dimensions of conservation increases, and capacity for government monitoring declines relative to an expanding network of protected areas, I suspect the central government will circle back to WMAs as a model that it views as a compromise between state-led fortress conservation and fully devolved models of community-based conservation.

Though Randilen's community members view the WMA as centrally important to local livelihoods, the institutional arrangements for governing it are not fully devolved. State policy dictates the legislative framework within which WMAs operate. Relevant acts are continually revised every few years, and representatives of the state must sign off on key processes prior to formalization. Tourism revenue is also centrally collected by TAWA before subsequently being distributed to the district level, then to the WMA, and finally to member villages. WMA management is supervised by Honeyguide, which is contracted directly by Randilen's Community-Based Organization. At first glance, these arrangements might lead skeptical social scientists to conclude that Randilen WMA is being run in a top-down manner constitutive of fortress conservation (Brehony et al. 2018). However, my central contention is that much more nuance is needed. While Randilen is constrained to some extent by the legislation governing all WMAs in the country, the actual institutions for governance are very much devolved to the community via their Authorized Association (AA). On key measures of governance, like how decisions are taken, communities are directly involved through AA meetings. I have participated in several of these gatherings and was struck by the extent to which Randilen WMA authorities listened to, and collaborated with, representatives from each village to take their concerns into consideration (chapter 5).

It became strikingly clear through my fieldwork that despite its complex beginnings, Randilen WMA has very much become a community-based conservation area. The head manager, accountant, WMA chair, and game scouts are all from the member villages. The WMA provides community members with a valuable dry season grazing bank, revenue for building schools and dispensaries, and vehicles to push elephants out of their farms in the wet season. Thus, I stand by its title as a community-based conservation area. The community still has some concerns, mainly to do with the livelihood impacts of human-wildlife interactions and the centralized nature in which WMA revenue is collected by TAWA before being dispersed back to the WMA—Randilen's residents would rather see this order of operations reversed—but overall, the community has come to support the WMA and view it as their own. With the help of Honeyguide, Randilen WMA has blossomed into a livelihood-oriented institution that community members are able to steer toward the things that matter to them: managing rangelands on a seasonal basis and defending farms from crop-raiding elephants. The result is a community-based conservation area that protects a key wildlife dispersal area adjacent to Tarangire National Park, while also providing a crucial grazing area for livestock, an effective crop-protection team for local farmers, and a considerable share of tourism revenue. Thus, Randilen WMA is a social enterprise that works for its community. The next step is helping the business reach financial independence, a task that Honeyguide has identified as a central component of its new strategic plan for the next five years.

Formalization and Pastoral Resilience

A crucial feature of Randilen WMA seems to be the way it formalizes seasonal grazing plans for its local pastoral community that previously existed through customary institutions. In this case, formal institutions of the state have melded with preexisting customary ones and, although they have transformed and adapted to a new political environment, they retain the same function. While not a direct application of the concept of "resilience" in ecological terms, I am springboarding from this concept to suggest that this synthesis between political policies from above and underlying pastoral institutions on the ground seems to reflect a somewhat resilient model of pasture management that achieves its own form of stability. Given the macropolitical context of wildlife conservation in Tanzania, WMAs are perhaps the best option available to pastoralists aiming to secure access to rangelands in wildlife dispersal areas that the state values for tourism.

The key challenge that will continually test the long-term resilience of this arrangement is drought, which substantially reduces available pasture for herders (cf. McCabe 1990). Randilen WMA has a stipulation that in such scenarios, the entire wildlife area can be used for livestock grazing, something which national parks do not formally have. Thus, while it is not a perfect solution that mirrors exactly the pastoral system of mobility from precolonial times, it could be an effective one even in the face of extreme events. The mettle of pastoral institutions is not measured in terms of productive maximization (Galaty and Johnson 1990), but by the capabilities of these arrangements to mitigate risk in the face of environmental uncertainty (McCabe 1997). Indeed, Randilen's resilience was put to the test during the severe drought of 2022, during which time it provided a crucial grazing bank for local pastoralists until the rains returned in March 2023.

An enduring issue from a resilience perspective is how WMAs function institutionally to regulate access of outsiders to pastures inside their territories during times of stress. Historically, drought has necessitated dissipation of traditional territorial boundaries in Maasailand (Galaty 2016). Unlike pastoral commons managed exclusively through customary institutions, however, WMAs are formalized by law, and their boundaries are strictly enforced against outsiders. During the 2022 drought, Randilen's AA exercised its authority to decide whether to allow herders from elsewhere to graze their livestock in the WMA. The AA superseded the sovereignty of individual village councils and the customary Maasai institutions that enable long-distance nomadic pastoralism. Through AA committee meetings, Randilen's community ultimately decided to open its drought reserve for *local* livestock keepers only, excluding herders from afar from benefiting from the WMA during a time of great stress. Thus, it must be acknowledged that WMAs have the potential to disrupt the pastoral commons through localization of rangeland management practices (cf. Lesorogol and Lesorogol 2024). This consideration highlights a potentially significant conservation trade-off for livelihood-oriented WMAs implemented in pastoral contexts—heightened benefits for local livestock keepers in the forms of land tenure security and enhanced range management capacity, but an associated risk of institutionafl fragmentation of the regional commons at a larger geographic scale.

The establishment of WMAs in Tanzania's pastoral areas transforms institutions for managing semi-arid rangelands, accelerating a process of formalization that began with land reforms in the colonial period and became entrenched through villagization. WMAs, like villages, create formal governance units recognized by state law that overlay existing customary patterns of common property resource access. As argued in this book, formalization has the potential to equip pastoral communities with new mechanisms for defending territory and protecting local livelihoods in the face of external actors at different scales who have interest in grabbing common lands for other purposes—wildlife conservation included. At the same time, it is also important to consider the broader geographic context of Maasailand and the crucial role that flexible boundaries and ethnic reciprocity play in ensuring resilience of the pastoral system in the face of extreme events like drought. Put back in a broader pastoral context, the very idea of tightly enforced local boundaries seems antithetical to the abilities of pastoralists to adapt to changing environmental conditions, which so often require movement across fixed boundaries and guarantees of hospitality when herders from one area arrive at another with their livestock in times of need. Historically, in such instances, Maasai have been welcoming to herders from afar, knowing that fortunes may at some point change, and those providing grass to graze in the present may find themselves at the doorsteps of others in the future when new hardships arise. Such networks of ethnic reciprocity spanning the varied landscapes of Maasailand across Kenya and Tanzania have been vital to the continuation of the pastoral way of life.

Political factors underpinning the fragmentation of common lands into private land holdings in Kenya and into jurisdictional villages in Tanzania seem to undermine the future prospects of pastoralism in the region. It is prudent to ask how the formalization of WMAs contributes to the wider institutional landscape of pastoral commons. WMAs—like villages—that are wielded by localities in their own interests at the expense of outside herders may indeed undercut pastoral resilience at a larger scale. Importantly, however, communities have agency in deciding how to use these institutions. The main issue affecting pastoral communities in northern Tanzania does not seem to be conflicts with other herders in times of stress, but in articulating tenure claims to the state, foreign investors, and encroaching smallholder cultivators. Formalization allows pastoralists to speak the language of the state while still allowing room on the ground for negotiation, interpretation, and flexibility. How laws are written on paper and how institutions are enacted in practice are often two very different things. In the case of Randilen, the WMA formalizes community tenure claims and protects the land from being grabbed for other purposes. The community retains the right to determine how the area is actively managed through their AA and considers flexible boundaries during times of stress to be crucial to the management model.

Given the history of political elites using communal grazing areas for their private benefit prior to the establishment of the WMA, the community was particular about using Randilen to prevent elites from taking advantage of them. This meant excluding livestock keepers from outside the community. Formalization also ensured that trophy hunters could not undermine their photographic tourism enterprise, that smallholder cultivators could not continue to encroach, and that absentee holders of commercial farms could be pushed off the land. Formalizing Randilen thus directly safeguarded the interests and well-being of its local community. Importantly, it also did not foreclose the future possibility of herders from elsewhere coming to Randilen in times of need. The WMA institution ensures that it is the local community, via the AA, that ultimately decides whether such visits will be permitted and under what conditions. Several

AA members and respected community elders have communicated to me that Randilen's AA would be willing to listen to the concerns of Maasai herders from other places and possibly grant grazing access during times of extreme drought, provided that the interactions were grounded in notions of respect and reciprocity and did not reduce the capacity of the WMA to sustain its local community.

It would seem, then, that if WMAs are governed and managed through participatory mechanisms, as Randilen is, then they could fit within the wider institutional landscape of pastoral commons insofar as they not only permit but *prioritize* livestock grazing as a central part of the management model and feature flexible boundaries to be enforced at the discretion of community members through the AA. WMAs that prohibit livestock grazing, or are governed in top-down ways that are not representative of community interests, are likely to create conflicts rather than foster cooperation. Blanket exclusions of pastoralists from WMAs, especially during times of stress, are a surefire way to undermine pastoral resilience and corrode local support for wildlife conservation on village land.

Key Takeaways

My conclusion is that although WMAs are not fully devolved to communities, they can reflect equitable and effective approaches to wildlife conservation on semi-arid rangelands. Randilen WMA has succeeded in garnering community-level support, albeit for complex reasons that are only tangentially related to the central objectives of conservation. Through the eyes of local pastoralists, conservation is less about wildlife than it is about *erematare*, or the overall form of rangeland management that operates through collective institutions (cf. Godfrey 2018). The key from an applied conservation perspective is ensuring that these primary concerns of pastoralists overlap with those of the central government and conservationists. Randilen WMA appears to accomplish this task by simultaneously securing large areas of wildlife habitat and formalizing institutions for rotational management of pasture, including the ever-essential grazing bank that provides a lifeline for local herders in the dry season. Randilen WMA now constitutes one of the largest formally managed grass banks in Tanzania, spanning 37,500 hectares of rangelands for community livestock grazing. Recent ecological studies also quantify rising numbers of wild mammals in Randilen WMA that are on par with those of Tarangire National Park (Lee and Bond 2018). These findings are highly encouraging and suggest that the ever-elusive goal of implementing conservation areas that simultaneously benefit people and wildlife is finally showing signs of attainability in Tanzania. Randilen's community has come to trust that conservation represents a set of practices that can support their livelihood goals rather than undermine them. Member villages of the WMA have come to reap the benefits from tourism revenue distributed to the village level. Residents of Lolkisale refer to these funds as "the milk of elephants," a phrase signifying the integration of wildlife conservation with their traditional way of life—milk of livestock being the economic dividends of the pastoral mode of production (cf. Lesorogol 2022). These benefits, while secondary to pasture access and range productivity, further reinforce community support for the WMA.

The result of these convergent social, political, and economic processes is that a significant transformation in community attitudes toward conservation has taken place in this part of Tanzania. Through formalization of institutional arrangements for governing and managing land involving some degree of decentralization, Randilen's community members have embraced conservation as an important aspect of their lives. These findings suggest that conservation in Tanzania does not have to protect wildlife habitat at the expense of excluding people. By including local communities in conservation governance and management, common ground can be reached between state, private, and community actors whose interests are diverse, but not always in direct competition with one another. Almost all community members I surveyed (94%; $n=634$) considered Randilen WMA a success rather than a failure, a glowing user experience review considering enduring complexities with the WMA model. Perhaps the prospect of community-based conservation in Tanzania is not a pipe dream after all.

ACKNOWLEDGMENTS

The research underpinning this book would not have been possible without the support of numerous institutions and people. Funding for doctoral fieldwork in 2019, 2020, and 2022 was provided by the Social Sciences and Humanities Research Council of Canada (SSHRC) and the Canadian International Development Research Centre (IDRC) via the Institutional Canopy of Conservation (ICAN) project (2019–2022), a Vanier Canada Graduate Scholarship (2018–2020), and a Michael Smith Foreign Study Supplement (2019). Dissertation field research was also enriched by a Salisbury Award from the Canadian Anthropology Society (2019–2020), a Graduate Research Award from the McGill Institute for the Study of International Development (ISID) (2018–2019), and a Schull-Yang International Experience Award (2018–2020).

Funding for field research in 2023 and 2024 was afforded by a Seed Grant from the Office of Research and Innovation Services at the University of Lethbridge (2024–2025), a University of Lethbridge Research Fund Grant (2023–2025), a SSHRC Explore Grant (2023–2024), a Research Affiliate Fund Grant from the Prentice Institute for Global Population and Economy (2023–2024), and a Start-Up Grant from the Faculty of Arts & Science at the University of Lethbridge (2023–2025). A SSHRC Exchange Grant (2023–2025) helped offset dissemination and publication costs.

Ethical reviews for the conduct of research with human subjects were administered by McGill University (479–0419) and the University of Lethbridge via the University of Alberta (Pro00130079; Pro00130079). Permits for carrying out research in Tanzania were obtained via the Tanzania Commission for Science and Technology (COSTECH) and the Tanzania Wildlife Research Institute (TAWIRI) (2019-426-NA-2019-299; CST00000398-2024-2024-00240). Permissions were also provided by Arusha and Manyara Regional Governments, Monduli and Babati District Councils, and Village Governments.

I am immensely grateful to Edwin Maingo Ole, Soipey Parkipuny, Malano Sinandei, and Melau Sinandei for research assistance and companionship throughout fieldwork. I want to acknowledge Elicia Bell for enhancing my understanding of wildlife dynamics through her community-based PhD research in Makuyuni village (2024), carried out with the capable help of Henry Emmanuel Leken. I would like to recognize Elsa DeLuca, Dakota Huddlestun, and George Tanner for assistance with interview transcriptions and Antoine Mounier for being a great flat mate during my doctoral studies in Montreal. I also want to thank Vinay Kamat for his continued mentorship and guidance, and for helping instill in me an appreciation for rigorous ethnographic methods.

I wish to acknowledge my NGO collaborators at the Ujamaa Community Resource Team (Makko Sinandei, Edward Loure), Pastoral Women's Council (Maanda Ngoitiko), Honeyguide (Damian Bell, Sam Shaba, Lemuta Meng'oru), the Nature Conservancy Tanzania (Chira Schouten, Alphonce Mallya, Lucas Yamat), Tanzania People and Wildlife (Loshiro Alais Morindat, Kirerenjo Medukenya), Tanzania Natural Resource Forum (Daniel Ouma), and the Kesho Trust (Emmanuel Ole Kileli). I want to thank Reuben Ole Kuney, Thadeus Clamian, Alais Morindat, and John Salehe for their insightful discussions and interviews about Maasailand and issues pertaining to pastoralism, wildlife conservation, and public policy. I am very grateful to the government staff at local, district, regional, and central levels who graciously facilitated my research and patiently participated as interviewees. In particular, I want to thank several members of the Monduli District Government (Seraphino Bichabicha Mawanja, Reginald Tesha, Catherine Francis Maembe) and government officials at the central level (Angello Joseph Mwilawa, Maurus Msuha, Stephen Nindi, Richard Kwitega). Richard, the Arusha regional secretary at the time of my doctoral fieldwork, died tragically in a head-on collision on the A104 highway in early 2021, and I will remember him for being patient and friendly and expressing genuine interest in the well-being of rural communities. I am also grateful to the representatives of the Tanzania Wildlife Management Authority (TAWA), the Tanzania National Parks Authority (TANAPA), and the Tanzania Wildlife Research Institute (TAWIRI) who participated in my study.

I owe gratitude to numerous local researchers for diversifying my study and broadening my analytical gaze. These include faculty members at the School for Field Studies in Karatu (Christian Kiffner, Bernard Kissui, John Mwamhanga), the College of African Wildlife Management, Mweka (Alex Kisingo, Kwaslema Malle Hariohay), Sokoine University of Agriculture (Agnes Sirima, George Mutani Msalya), the Tarangire Elephant Project (Boniface Osujaki), and the Tarangire Lion Project (Leo Mwanga). I also enjoyed insightful conversations while in the field with Jim Igoe, Monica Bond, Derek Lee, Peadar Brehony, Alicia Davis, Tim Davenport, and Monique Borgerhoff Mulder. I feel indebted to the scholars whose publications were vital for linking up my empirical study with the wider historical context of conservation in Tanzania's Maasailand. These include Dorothy Hodgson, Benjamin Gardner, Helge Kjekshus, Fred Nelson, Thomas Spear, Katherine Homewood, Terrence McCabe, Mara Goldman, Dan Brockington, Stephanie Loveless, Jevgeniy Bluwstein, Daniel Ndagala, Roderick Neumann, Hassan Sachedina, Emmanuel Sulle, and Olivier LaRocque. I hope this book contributes to an already vibrant discourse community.

I want to acknowledge the manager of Randilen Wildlife Management Area (WMA), Meshurie Melembuki, and the Randilen WMA chair at the time of my doctoral fieldwork, Kilamian Lendoya, who were gracious in supporting my research and who have been working hard to foster an effective and equitable model of community-based conservation outside Tarangire National Park. I want to extend this thank you to the village game scouts of Randilen WMA who have been striving to uphold pastoral values and safeguard wildlife habitat. I am especially grateful to them for hosting me at their ranger posts and participating in interviews. I want to recognize Lotha Lukas Nooi, Aloyce

Alphonce Mjengi, and Michael Melau for field assistance and guidance in the study villages. I also want to thank the village governments and assemblies of each village for permitting me to work in their communities. I express a heartfelt thank you in particular to those families who welcomed me into their homes and gave me a glimpse of the joys and challenges that animate their everyday lives.

During my doctoral studies (2016–2022), I benefited from the expertise of the faculty in the Department of Anthropology at McGill University, particularly Colin Scott and Ismael Vaccaro, who served on my PhD committee. I want to especially thank my doctoral supervisor, John Galaty. John and I started working together on the ICAN project nearly a decade ago, and he has been unwaveringly patient and supportive of my research program since then. He has shared with me a wealth of knowledge about the Maasai, East African historiography, and the social-ecological dynamics of pastoralism. Perhaps most significantly, he has always treated me with respect and kindness. Through the ICAN project, John generated institutional backing for my research that was crucial for carrying out my doctoral fieldwork and cultivating a social network in East Africa. In the context of my PhD oral defense, Sandra Hyde, Terrence McCabe, and Jon Unruh posed thoughtful and constructive questions that helped refine the analysis that informs this book. As well, I appreciate the social capital generated by my former ICAN graduate student colleagues, particularly Kariuki Kirigia, Corey Wright, Stephen Moiko, Klerkson Lugusa, Graham Fox, Kathleen Godfrey, and Salau Rogei.

Some of the theoretical lines of argumentation in this book developed during a visiting postdoctoral fellowship (2022) at Harvard University's Program on Science, Technology and Society (STS) through interactions with other fellows and faculty members. I am grateful to the STS community at Harvard for helping me explore the STS currents in my work. I would also like to acknowledge the network of academic institutions I am currently affiliated with, including the Prentice Institute for Global Population and Economy, the Centre for Indigenous Conservation and Development Alternatives (CICADA), the McGill Centre for Society, Technology and Development (STANDD), the Indian Ocean World Centre (IOWC), and the Applied Research in Environmental Anthropology (AREA) Lab. I am fortunate to have great colleagues in the Department of Anthropology at the University of Lethbridge who contribute regularly to my intellectual life as an anthropologist.

With regard to the publication process, the University of Georgia Press has been a delight to work with as an author. A special word of thanks is in order for my acquisitions editor, Mick Gusinde-Duffy, who shepherded me through the book publication process with warmth, encouragement, and professionalism. I am thankful for the support of the *Geographies of Justice and Social Transformation* team, including the series editors Matt Coleman and Ishan Ashutosh, who helped ensure the book was attuned to the series's readership, and to Bethany Snead, Jon Davies, and Michelle Witkowski (Westchester Publishing) for keeping the publication process moving behind the scenes. Two anonymous readers contributed greatly to the development of this book through the provision of constructive peer reviews that helped sharpen my arguments and situate the study in relation to other works. Michael Goldstein adeptly prepared the index.

Lastly, I have been lucky to have a dedicated family behind me as I embarked on fieldwork across the globe and spent countless holidays glued to the screen of my laptop. I am deeply thankful to my parents, Niloofer Baria and Stephen Raycraft, for instilling in me at a young age the value of education; to my brothers, Cody Raycraft and Tyler Raycraft, for being there for me; and to my in-laws, Marion Anderson, Dennis Ejack, and Stephen Ejack, for helping me feel at home in Alberta. Most of all, I want to thank my wife, Leanne Ejack, for furnishing me with immeasurable emotional support, for keeping me healthy and organized, and for providing me with the love that makes all of life's pursuits worthwhile.

REFERENCES

Agrawal, Arun. 2005. *Environmentality: Technologies of government and the making of subjects.* Durham, N.C.: Duke University Press.

Agrawal, Nisha, Zafar Ahmed, Michael Mered, and Roger Nord. 1993. *Structural adjustment, economic performance, and aid dependency in Tanzania.* Washington, D.C.: World Bank.

Arlin, Camilla. 2011. "Becoming wilderness: A topological study of Tarangire, northern Tanzania, 1890–2000." PhD Dissertation. Department of Human Geography. Stockholm University.

Baldus, Rolf D., and Andrew E. Cauldwell. 2004. "Tourist Hunting and its Role in Development of Wildlife Management Areas in Tanzania." Sixth International Game Ranching Symposium, Paris, July 6–9, 2004.

Barkan, Joel D. 1994. *Beyond capitalism vs. socialism in Kenya and Tanzania.* London, UK: Lynne Rienner Publishers.

Behnke, Roy H., and Ian Scoones. 1993. "Rethinking range ecology: Implications for rangeland management in Africa." In *Range ecology at disequilibrium: New models of natural variability and pastoral adaptation in African savannas,* edited by Roy H. Behnke, Ian Scoones and Carol Kerven, 1–30. London, UK: Overseas Development Institute.

Benjaminsen, Tor A., and Ian Bryceson. 2012. "Conservation, green/blue grabbing and accumulation by dispossession in Tanzania." *Journal of Peasant Studies* 39(2): 335–355.

Benjaminsen, Tor A., Mara J. Goldman, Maya Y. Minwary, and Faustin P. Maganga. 2013. "Wildlife management in Tanzania: State control, rent seeking and community resistance." *Development and Change* 44(5): 1087–1109.

Bennett, Nathan J. 2015. "Governing marine protected areas in an interconnected and changing world." *Conservation Biology* 29(1): 303–306.

Bennett, Nathan James, and Philip Dearden. 2014a. "From measuring outcomes to providing inputs: Governance, management, and local development for more effective marine protected areas." *Marine Policy* 50: 96–110.

———. 2014b. "Why local people do not support conservation: Community perceptions of marine protected area livelihood impacts, governance and management in Thailand." *Marine Policy* 107–116.

Bennett, Nathan J., Robin Roth, Sarah C. Klain, Kai Chan, Patrick Christie, Douglas A. Clark, Georgina Cullman, Deborah Curran, Trevor J. Durbin, and Graham Epstein. 2017. "Conservation social science: Understanding and integrating human dimensions to improve conservation." *Biological Conservation* 205: 93–108.

Bennett, Nathan J., and Terre Satterfield. 2018. "Environmental governance: A practical framework to guide design, evaluation, and analysis." *Conservation Letters* 11(6): 1–13.

Bluwstein, Jevgeniy. 2017. "Creating ecotourism territories: Environmentalities in Tanzania's community-based conservation." *Geoforum* 101–113.

———. 2018a. "Biopolitical landscapes: Governing people and spaces through conservation in Tanzania." PhD Dissertation. Department of Food and Resource Economics. University of Copenhagen.

———. 2018b. "From colonial fortresses to neoliberal landscapes in northern Tanzania: A biopolitical ecology of wildlife conservation." *Journal of Political Ecology* 25(1): 144–168.

———. 2019. "Resisting legibility: State and conservation boundaries, pastoralism, and the risk of dispossession through geospatial surveys in Tanzania." *Rural Landscapes: Society, Environment, History* 6(1):1–18.

———. 2022. "Historical political ecology of the Tarangire Ecosystem: From Colonial Legacies, to Contested Histories, Towards Convivial Conservation?" In *Tarangire: Human-wildlife coexistence in a fragmented ecosystem*, edited by Christian Kiffner, Monica L. Bond and Derek E. Lee, 25–46. Cham, Switzerland: Springer.

Bluwstein, Jevgeniy, and Jens Friis Lund. 2018. "Territoriality by conservation in the Selous-Niassa Corridor in Tanzania." *World Development* 101: 453–465.

Bluwstein, Jevgeniy, Francis Moyo, and Rose Peter Kicheleri. 2016. "Austere conservation: Understanding conflicts over resource governance in Tanzanian wildlife management areas." *Conservation & Society* 14(3): 218–231.

Bond, Monica L., Derek E. Lee, and Christian Kiffner. 2022. "Towards human-wildlife coexistence in the Tarangire ecosystem." In *Tarangire: Human-Wildlife Coexistence in a Fragmented Ecosystem*, edited by Christian Kiffner, Monica L. Bond and Derek E. Lee, 367–391. Cham, Switzerland: Springer.

Borner, Markus. 1982. *Recommendations for a multiple land use authority adjacent to Tarangire National Park, Arusha Region, Tanzania*. Arusha: Frankfurt Zoological Society.

———. 1985. "The increasing isolation of Tarangire National Park." *Oryx* 19(2) 91–96.

Brehony, Peadar, Jevgeniy Bluwstein, Jens Friis Lund, and Peter Tyrrell. 2018. "Bringing back complex socio-ecological realities to the study of CBNRM impacts: A response to Lee and Bond (2018)." *Journal of Mammalogy* 99(6): 1539–1542.

Brehony, Peadar, Alais Morindat, and Makko Sinandei. 2022. "Land tenure, livelihoods, and conservation: Perspectives on priorities in Tanzania's Tarangire Ecosystem." In *Tarangire: Human wildlife coexistence in a fragmented ecosystem*, edited by Christian Kiffner, Monica L. Bond and Derek E. Lee, 85–108. Cham, Switzerland: Springer.

Brockingon, Dan. 1999. "Conservation, displacement and livelihoods. The consequences of the eviction for pastoralists moved from the Mkomazi Game Reserve, Tanzania." *Nomadic Peoples* 3(2): 74–96.

———. 2002. *Fortress conservation: The preservation of the Mkomazi Game Reserve, Tanzania*. Bloomington, Ind.: Indiana University Press.

———. 2008. "Preserving the New Tanzania: Conservation and land use change." *International Journal of African Historical Studies* 41(3): 557–579.
Brockington, Dan, Rosaleen Duffy, and Jim Igoe. 2012. *Nature unbound: Conservation, capitalism and the future of protected areas*. London, UK: Routledge.
Brockington, Dan, and James Igoe. 2006. "Eviction for conservation: A global overview." *Conservation & Society* 4(3): 424.
Bromley, Daniel W. 1992. "The commons, common property, and environmental policy." *Environmental and Resource Economics* 2(1): 1–17.
Bryant, Raymond L., and Sinead Bailey. 1997. *Third world political ecology*. New York, N.Y.: Routledge.
Büscher, Bram, and Veronica Davidov. 2013. *The ecotourism-extraction nexus: Political economies and rural realities of (un) comfortable bedfellows*. New York, N.Y.: Routledge.
Büscher, Bram, and Robert Fletcher. 2015. "Accumulation by conservation." *New Political Economy* 20(2): 273–298.
———. 2019. "Towards convivial conservation." *Conservation & Society* 17(3): 283–296.
Büscher, Bram, Sian Sullivan, Katja Neves, Jim Igoe, and Dan Brockington. 2012. "Towards a synthesized critique of neoliberal biodiversity conservation." *Capitalism Nature Socialism* 23(2): 4–30.
Caro, Tim, and Tim R. B. Davenport. 2016. "Wildlife and wildlife management in Tanzania." *Conservation Biology* 30(4): 716–723.
Cepek, Michael L. 2011. "Foucault in the forest: Questioning environmentality in Amazonia." *American Ethnologist* 38: 501–515.
Cleaver, Frances. 2012. *Development through bricolage: Rethinking institutions for natural resource management*. London, UK: Routledge.
Cochran, William G. 1963. *Sampling techniques*. 2nd ed. New York, N.Y.: John Wiley and Sons.
Cronon, William. 1996. "The trouble with wilderness: Or, getting back to the wrong nature." *Environmental History* 1(1): 7–28.
Davis, Alicia. 2011. "'Ha! What is the benefit of living next to the park?' Factors limiting in-migration next to Tarangire National Park, Tanzania." *Conservation & Society* 9(1): 25.
de Sardan, Jean-Pierre Oliver. 2005. *Anthropology and development: Understanding contemporary social change*. London, UK: Zed Books.
de Soto, Hernando. 2000. *The mystery of capital: Why capitalism triumphs in the West and fails everywhere else*. New York, N.Y.: Basic Books.
Descola, Philippe. 2013. *Beyond nature and culture*. Chicago, Ill.: University of Chicago Press.
Ferguson, James. 1994. *The anti-politics machine: "Development," depoliticization, and bureaucratic power in Lesotho*. Minneapolis, Minn.: University of Minnesota Press.
Fimbo, Gamaliel Mgongo. 1992. *Essays in land law, Tanzania*. Dar es Salaam, Tanzania: Faculty of Law, University of Dar es Salaam.
Fletcher, Robert. 2017. "Environmentality unbound: Multiple governmentalities in environmental politics." *Geoforum* 85: 311–315.
———. 2023. *Failing forward: The rise and fall of neoliberal conservation*. Berkeley, Calif.: University of California Press.

Fletcher, Robert, Wolfram H. Dressler, Zachary R. Anderson, and Bram Büscher. 2018. "Natural capital must be defended: Green growth as neoliberal biopolitics." *Journal of Peasant Studies* 46(5): 1–28.

Fletcher, Robert, Wolfram Dressler, and Bram Büscher. 2014. "The new frontiers of environmental conservation." In *Nature™ Inc.*, edited by Bram Büscher, Wolfram Dressler and Robert Fletcher, 3–24. Tucson, Ariz.: University of Arizona Press.

Fodor's Travel Guide (Fodor's). 2006. "Post #15." Accessed Nov. 15, 2024. https://www.fodors.com/community/africa-and-the-middle-east/eastco-trip-574514/.

Foley, Charles A. H., and Lisa J. Faust. 2010. "Rapid population growth in an elephant Loxodonta africana population recovering from poaching in Tarangire National Park, Tanzania." *Oryx* 44(2): 205–212.

Foley, Charles A. H., and Lara S. Foley. 2022. "The history, status, and conservation of the elephant population in the Tarangire ecosystem." In *Tarangire: Human-wildlife coexistence in a fragmented ecosystem*, edited by Christian Kiffner, Monica L. Bond and Derek E. Lee, 209–232. Cham, Switzerland: Springer.

Fosbrooke, Henry A. 1948. "An administrative survey of the Masai social system." *Tanganyika Notes and Records* 26: 1–50.

———. 1956a. "'Introduction and Annotations,' to 'The Life of Justin: An African Autobiography' by Justin Lemenye." *Tanganyika Notes and Records* 41: 31–57.

———. 1956b. "The Masai age-group system as a guide to tribal chronology: Part I: The Masai system." *African Studies* 15(4): 188–206.

Foucault, Michel. 1976. *"Society must be defended" Lectures at the College de France 1975–76*. Translated by David Macey. Edited by Mauro Bertani and Alessandro Fontana. New York, N.Y.: Picador.

———. 1978. *The history of sexuality, volume I*. New York, N.Y.: Vintage Books.

Galaty, John G. 1982. "Being 'Maasai'; being 'people-of-cattle': Ethnic shifters in East Africa." *American Ethnologist* 9(1): 1–20.

———. 1993. "Maasai expansion and the New East African pastoralism." In *Being Maasai? Ethnicity & identity in East Africa*, edited by Thomas Spear and Richard Waller, 61–86. London, UK: James Currey.

———. 1994. "Ha(l)ving land in common: The subdivision of Maasai group ranches in Kenya." *Nomadic Peoples* 34/35: 109–122.

———. 2016. "Reasserting the commons: Pastoral contestations of private and state lands in East Africa." *International Journal of the Commons* 10(2): 1–19.

Galaty, John G., and D. Johnson. 1990. *The world of pastoralism: Herding systems in comparative perspective*. New York, N.Y.: Guildford Press.

Gardner, Benjamin. 2007. "Producing pastoral power: Territory, identity and rule in Tanzanian Maasailand." PhD Dissertation. Department of Geography. University of California, Berkeley.

———. 2012. "Tourism and the politics of the global land grab in Tanzania: Markets, appropriation and recognition." *Journal of Peasant Studies* 39(2): 377–402.

———. 2016. *Selling the Serengeti: The cultural politics of safari tourism*. Athens, Ga.: University of Georgia Press.

Gibbon, Peter, and Philip L. Raikes. 1995. *Structural adjustment in Tanzania, 1986–94*. Copenhagen, Denmark: Danish Ministry of Foreign Affairs.

Godfrey, Kathleen B. H. 2018. "Toward erematare, beyond conservation: Meaning, practice, and rethinking the conservation story in the Maasai communities of Olkiramatian and Shompole, Kajiado County, Kenya." Master's thesis. Department of Anthropology. McGill University.

Goldman, Mara J. 2003. "Partitioned nature, privileged knowledge: Community-based conservation in Tanzania." *Development and Change* 34(5): 833–862.

———. 2009. "Constructing connectivity: Conservation corridors and conservation politics in East African rangelands." *Annals of the Association of American Geographers* 99(2): 335–359.

———. 2011. "Strangers in their own land: Maasai and wildlife conservation in northern Tanzania." *Conservation & Society* 9(1): 65.

———. 2018. "Circulating wildlife: Capturing the complexity of wildlife movements in the Tarangire ecosystem in northern Tanzania from a mixed method, multiply situated perspective." In *The Palgrave handbook of critical physical geography*, edited by Rebecca Lave, Christine Biermann and Stuart N. Lane, 319–338. New York, N.Y.: Springer.

———. 2020. *Narrating nature: Wildlife conservation and Maasai ways of knowing*. Tucson, Ariz.: University of Arizona Press.

Green, Kathryn E., and William M. Adams. 2015. "Green grabbing and the dynamics of local-level engagement with neoliberalization in Tanzania's wildlife management areas." *Journal of Peasant Studies* 42(1): 97–117.

Hardin, Garrett. 1968. "The tragedy of the commons: The population problem has no technical solution; it requires a fundamental extension in morality." *Science* 162(3859): 1243–1248.

———. 1994. "The tragedy of the unmanaged commons." *Trends in Ecology and Evolution* 9(5): 199.

Hariohay, Kwaslema Malle, and Eivin Røskaft. 2015. "Wildlife induced damage to crops and livestock loss and how they affect human attitudes in the Kwakuchinja Wildlife Corridor in northern Tanzania." *Environment and Natural Resources Research* 5(3): 56–63.

Havnevik, Kjell J. 1993. *Tanzania: The limits to development from above*. Dar es Salaam, Tanzania: Nordic Africa Institute.

Hodgson, Dorothy Louise. 2001. *Once intrepid warriors: Gender, ethnicity, and the cultural politics of Maasai development*. Bloomington, Ind.: Indiana University Press.

Hoffman, David M. 2014. "Conch, cooperatives, and conflict: Conservation and resistance in the Banco Chinchorro Biosphere Reserve." *Conservation & Society* 12(2): 120–132.

Holmes, George. 2007. "Protection, politics and protest: Understanding resistance to conservation." *Conservation & Society* 5(2): 184–201.

Holmes, George, and Connor J. Cavanagh. 2016. "A Review of the Social Impacts of Neoliberal Conservation: Formations, Inequalities, Contestations." *Geoforum* 75: 199–209.

Homewood, Katherine. 1994. "Range ecology at disequilibrium: New models of natural variability and pastoral adaptation in African savannas." *Africa: Journal of the International African Institute* 64(4): 581–583.

———. 1995. "Development, demarcation and ecological outcomes in Maasailand." *Africa* 65(3): 331–350.

———. 2008. *Ecology of African pastoralist societies.* Melton, UK: James Currey.

Homewood, Katherine, Jevgeniy Bluwstein, Jens Friis Lund, Aidan Keane, Martin R. Nielsen, Maurus Msuha, Joseph Olila, and Neil Burgess. 2015. *The economic and social viability of Tanzanian wildlife management areas.* Copenhagen, Denmark: University of Copenhagen.

Homewood, Katherine, Ernestina Coast, and Michael Thompson. 2004. "In-migrants and exclusion in East African rangelands: Access, tenure and conflict." *Africa* 74(4): 567–610.

Homewood, Katherine, Patti Kristjanson, and Pippa C. Trench. 2009. "Changing land use, livelihoods and wildlife conservation in Maasailand." In *Staying Maasai? Livelihoods, conservation and development in East African rangelands*, edited by Katherine Homewood, Patti Kristjanson and Pippa C. Trench, 1–42. New York, N.Y.: Springer.

Homewood, Katherine, Martin Reinhardt Nielsen, and Aidan Keane. 2020. "Women, wellbeing and wildlife management areas in Tanzania." *Journal of Peasant Studies*: 1–28.

Homewood, Katherine, and William A. Rodgers. 1987. "Pastoralism, conservation and the overgrazing controversy." In *Conservation in Africa: People, policies and practice*, edited by David Anderson and Richard Grove, 111–128. Cambridge, UK: Cambridge University Press.

Homewood, Katherine, and William A. Rodgers. 1991. *Maasailand ecology: Pastoralist development and wildlife conservation in Ngorongoro, Tanzania.* Cambridge, UK: Cambridge University Press.

Homewood, Katherine, and Michael Thompson. 2010. "Social and economic challenges for conservation in East African rangelands: Land use, livelihoods and wildlife change in Maasailand." In *Wild rangelands: Conserving wildlife while maintaining livestock in semi-arid ecosystems*, edited by Johan T Du Toit, Richard Kock and James Deutsch, 340–366. Hoboken, N.J.: John Wiley & Sons.

Homewood, Katherine, Pippa C. Trench, and Dan Brockington. 2012. "Pastoralism and conservation—who benefits?" In *Biodiversity conservation and poverty alleviation: Exploring the evidence for a link*, edited by Dilys Rose, Joanna Elliott, Chris Sandbrook and Matt Walpole, 239–252. Hoboken, N.J.: John Wiley & Sons.

Honeyguide. 2017. *Strategic plan 2017–2021.* Arusha, Tanzania: Honeyguide Foundation.

Huizer, Gerrit. 1973. "The Ujamaa village programme in Tanzania: New forms of rural development." *Studies in Comparative International Development* 8: 183–207.

Igoe, Jim. 2004. *Conservation and globalization: A study of national parks and indigenous communities from East Africa to South Dakota.* Belmont, Calif.: Wadsworth Publishing.

———. 2010. "The spectacle of nature in the global economy of appearances: Anthropological engagements with the spectacular mediations of transnational conservation." *Critique of Anthropology* 30(4): 375–397.

———. 2017. *The nature of the spectacle: On images, money, and conserving capitalism.* Tucson, Ariz.: University of Arizona Press.

———. 2022. "A conservationist political ecology in and for the Tarangire ecosystem." In *Tarangire: Human-wildlife coexistence in a fragmented ecosystem*, edited by Christian Kiffner, Monica L. Bond and Derek E. Lee, 47–63. Cham, Switzerland: Springer.
Igoe, Jim, and Dan Brockington. 1999. *Pastoral land tenure and community conservation: A case study from North-East Tanzania*. London, UK: International Institute for environment and development (IIED).
———. 2007. "Neoliberal conservation: A brief introduction." *Conservation & Society* 5(4): 432–449.
Igoe, Jim, and Beth Croucher. 2007. "Conservation, commerce, and communities: The story of community-based wildlife management areas in Tanzania's northern tourist circuit." *Conservation & Society* 5(4): 534–561.
Jackson, Jonathan M. 2021. "'Off to Sugar Valley': The Kilombero settlement scheme and 'Nyerere's People', 1959–69." *Journal of Eastern African Studies* 15(3): 505–526.
Jasanoff, Sheila. 2004. *States of knowledge: The co-production of science and the social order*. New York, N.Y.: Routledge.
Jones, Samantha. 2006. "A political ecology of wildlife conservation in Africa." *Review of African Political Economy* 33(109): 483–495.
Kamat, Vinay R. 2024. *In a wounded land: Conservation, extraction, and human well-being in coastal Tanzania*. Tucson, Ariz.: University of Arizona Press.
Keane, Aidan, Jens Friis Lund, Jevgeniy Bluwstein, Neil D. Burgess, Martin Reinhardt Nielsen, and Katherine Homewood. 2020. "Impact of Tanzania's wildlife management areas on household wealth." *Nature Sustainability* 3(3): 226–233.
Kent, Evelyn L., and Laly Lichtenfeld. 2024. "Case study: Big cats in the Maasai Steppe." Accessed Nov. 18, 2024. https://education.nationalgeographic.org/resource/case-study-big-cats-maasai-steppe/.
Kicheleri, Rose P. 2018. "Dispossession and power struggles in community-based natural resources management: A case of Burunge Wildlife Management Area, Tanzania." PhD Dissertation. Department of Forest Resources Assessment and Management. Sokoine University of Agriculture.
Kicheleri, Rose P., Lazaro J. Mangewa, Martin R. Nielsen, George C. Kajembe, and Thorsten Treue. 2021. "Designed for accumulation by dispossession: An analysis of Tanzania's wildlife management areas through the case of Burunge." *Conservation Science and Practice* 3(1): 1–15.
Kiffner, Christian, Monica L. Bond, and Derek E. Lee. 2022. "Human-wildlife interactions in the Tarangire ecosystem." In *Tarangire: Human-wildlife coexistence in a fragmented ecosystem*, edited by Christian Kiffner, Monica L. Bond and Derek E. Lee, 3–22. Cham, Switzerland: Springer.
Kiffner, Christian, Charles A. H. Foley, Derek E. Lee, Monica L. Bond, John Kioko, Bernard M. Kissui, Alex L. Lobora, Lara S. Foley, and Fred Nelson. 2024. "The contribution of community-based conservation models to conserving large herbivore populations." *Scientific Reports* 14(1):16221.
Kileli, Emmanuel Ole. 2017. "Assessment of the effectiveness of a community-based conservation approach used by pastoralist villages in Loliondo Division, northern Tanzania." Master's thesis. Department of Geography. University of Victoria.

Kimario, Fidelcastor, Nina Botha, Alex Kisingo, and Hubert Job. 2020. "Theory and practice of conservancies: Evidence from wildlife management areas in Tanzania." *Erdkunde* 77(2): 117–141.

King, Simon J. 2009. *Tarangire conservation area management plan revised October 2009*. Arusha, Tanzania: TCA.

Kipuri, Naomi, and Benedict Nangoro. 1996. *Community benefits through wildlife resources: Evaluation report for TANAPA's community conservation service programme*. Arusha, Tanzania: TANAPA.

Kissui, Bernard M., Elvis L. Kisimir, Laly L. Lichtenfeld, Elizabeth M. Naro, Robert A. Montgomery, and Christian Kiffner. 2022. "Human-carnivore coexistence in the Tarangire ecosystem." In *Tarangire: Human-wildlife coexistence in a fragmented ecosystem*, edited by C. Kiffner, Monica Bond and Derek Lee, 295–317. Cham, Switzerland: Springer.

Kjekshus, Helge. 1977. "The Tanzanian villagization policy: Implementational lessons and ecological dimensions." *Canadian Journal of African Studies* 11(2): 269–282.

Kuney, Reuben Ole. 1994. "Pluralism and ethnic conflict in Tanzania's arid lands: The case of the Maasai and the WaArusha." *Nomadic Peoples*: 95–107.

LaRocque, Olivier. 2006. "'The land is getting smaller': Changing territorial strategies of pastoralists in Tanzania." Master's thesis. Department of Anthropology. Mcgill University.

Leader-Williams, Nigel. 2000. "The effects of a century of policy and legal change on wildlife conservation and utilisation in Tanzania." In *Wildlife conservation by sustainable use*, edited by Herbert H. T. Prins, Jan G. Grootenhuis and Thomas T. Dolan, 219–245. Boston, Mass.: Kluwer Academic Publishers.

Lee, Derek E., and Monica L. Bond. 2018. "Quantifying the ecological success of a community-based wildlife conservation area in Tanzania." *Journal of Mammalogy* 99(2): 459–464.

Lemke, Thomas. 2001. "'The Birth of bio-Politics': Michel Foucault's lecture at the Collège de France on neo-liberal governmentality." *Economy and Society* 30(2): 190–207.

———. 2002. "Foucault, governmentality, and critique." *Rethinking Marxism* 14(3): 49–64.

Lesorogol, Carolyn K. 2022. *Conservation and community in Kenya: Milking the elephant*. Lanham, Md.: Lexington Books.

Lesorogol, Carolyn K., and Prame Lesorogol. 2024. "Community-based wildlife conservation on pastoral lands in Kenya: A new logic of production with implications for the future of pastoralism." *Human Ecology* 52(1): 15–29

Li, Tania M. 2007. *The Will to Improve: Governmentality, Development, and the Practice of Politics*. Durham, N.C.: Duke University Press.

Lolkisale Biodiversity Conservation Support Project (LBCSP). 2003. *A medium sized project brief submitted to the global enivronment facility*. Washington, D.C.: Global Environmental Facility.

Loveless, Stephanie. 2014. "Establishing WMAs in Tanzania: The role of community-level participation in the making of Randileni WMA." M.Sc thesis. Department of Food and Resource Economics. University of Copenhagen.

Mbaria, John, and Mordecai Ogada. 2016. *The big conservation lie: The untold story of wildlife conservation in Kenya*. Springfield, Mo.: Lens & Pens.

McCabe, J. Terrence. 1990. "Success and failure: The breakdown of traditional drought coping institutions among the pastoral Turkana of Kenya." *Journal of Asian and African studies* 25(3–4): 146–160.

———. 1997. "Risk and uncertainty among the Maasai of the Ngorongoro conservation area in Tanzania: A case study in economic change." *Nomadic Peoples*: 54–65.

———. 2003a. "Disequilibrial ecosystems and livelihood diversification among the Maasai of northern Tanzania: Implications for conservation policy in eastern Africa." *Nomadic Peoples* 7(1): 74–91.

———. 2003b. "Sustainability and livelihood diversification among the Maasai of northern Tanzania." *Human Organization* 62(2): 100–111.

McCabe, J. Terrence, Paul W. Leslie, and Alicia Davis. 2020. "The emergence of the village and the transformation of traditional institutions: A case study from northern Tanzania." *Human Organization* 79(2): 150–160.

McCabe, J. Terrence, Paul W. Leslie, and Laura DeLuca. 2010. "Adopting cultivation to remain pastoralists: The diversification of Maasai livelihoods in northern Tanzania." *Human Ecology* 38(3): 321–334.

McCabe, J. Terrence, and Emily Woodhouse. 2022. "Maasai wellbeing and implications for wildlife migrating from Tarangire National Park." In *Tarangire: Human-wildlife coexistence in a fragmented ecosystem*, edited by Christian Kiffner, Monica Bond and Derek Lee, 65–84. Cham, Switzerland: Springer.

McShane, Thomas O., Paul D. Hirsch, Tran Chi Trung, Alexander N. Songorwa, Ann Kinzig, Bruno Monteferri, David Mutekanga, et al. 2011. "Hard choices: Making trade-offs between biodiversity conservation and human well-being." *Biological Conservation* 144(3): 966–972.

Ministry of Natural Resources and Tourism (MNRT), Tanzania. 1998. *The wildlife policy of Tanzania*. Dodoma, Tanzania: MNRT.

———. 2009. *The Wildlife Conservation Act No. 5*. Dodoma, Tanzania: MNRT.

Monduli District Council (MDC), Tanzania. 1994. *Environmental profile Monduli district*. Monduli, Tanzania: MDC.

Moore, Donald S. 1998. "Subaltern struggles and the politics of place: Remapping resistance in Zimbabwe's eastern highlands." *Cultural Anthropology* 13(3): 344–81.

Moyo, Francis, Jasper Ijumba, and Jens Friis Lund. 2016. "Failure by design? Revisiting Tanzania's flagship wildlife management area Burunge." *Conservation & Society* 14(3): 232–242.

Msoffe, Fortunata U., Shem C. Kifugo, Mohammed Y. Said, Moses Ole Neselle, Paul Van Gardingen, Robin S. Reid, Joseph O. Ogutu, et al. 2011. "Drivers and impacts of land-use change in the Maasai Steppe of northern Tanzania: An ecological, social and political analysis." *Journal of Land Use Science* 6(4): 261–281.

Mulrennan, Monica E., Colin H. Scott, and Katherine Scott. 2019. *Caring for Eeyou Istchee: Protected area creation on Wemindji Cree territory*. Vancouver, BC: UBC Press.

Nash, Roderick Frazier. 2014. *Wilderness and the American mind*. New Haven, Conn.: Yale University Press.

Ndagala, Daniel K. 1982. "'Operation Imparnati': The sedentarization of the pastoral Maasai in Tanzania." *Nomadic Peoples* (10): 28–39.

———. 1994. "Pastoral territory and policy debates in Tanzania." *Nomadic Peoples* (34/35): 23–36.

———. 1997. "Pastoral resource access and control and the desertification process in East Africa." Atelier sur le foncier et la desertification, Dakar, Senegal, March 7–9, 1994.

Nelson, Fred. 2003. *Community-based tourism in northern Tanzania: Increasing opportunities, escalating conflicts and an uncertain future*. Amsterdam, Netherlands: Rozenberg.

———. 2004. *The evolution and impacts of community-based ecotourism in northern Tanzania*. Vol. 131. London, UK: IIED.

———. 2005. *Wildlife management and village land tenure in northern Tanzania*. Land Symposium, Dar es Salaam, March 1–2, 2005. Arusha, Tanzania: Tanzania Natural Resource Forum (TNRF).

Nelson, Fred, Charles Foley, Lara S. Foley, Abraham Leposo, Edward Loure, David Peterson, Mike Peterson, et al. 2010. "Payments for ecosystem services as a framework for community-based conservation in northern Tanzania." *Conservation Biology* 24(1): 78–85.

Nelson, Fred, Rugemeleza Nshala, and W. A. Rodgers. 2007. "The evolution and reform of Tanzanian wildlife management." *Conservation & Society* 5(2): 232–261.

Nelson, Fred, Emmanuel Sulle, and P. Ndoipo. 2006. *WMAs in Tanzania: A status report and interim evaluation*. Arusha, Tanzania: TNRF.

Neumann, Roderick P. 1995. "Local challenges to global agendas: Conservation, economic liberalization and the pastoralists' rights movement in Tanzania." *Antipode* 27(4): 363–382.

———. 1998. *Imposing wilderness: Struggles over livelihood and nature preservation in Africa*. Vol. 4. Berkeley, Calif.: University of California Press.

North, Douglass C. 1991. "Institutions." *Journal of Economic Perspectives* 5(1): 97–112.

Northern Tanzania Rangelands Initiative (NTRI). 2017. *Making wildlife management areas deliver for conservation and communities*. Arusha, Tanzania: NTRI.

Nshala, Rugemeleza, Vincent Shauri, Tundu Lissu, Bwire Kwaare, and Simon Metcalfe. 1998. *Socio-legal analysis of community based conservation in Tanzania: Policy, legal, institutional and programmatic issues, considerations and options*. Dar es Salaam, Tanzania: United States Agency for International Development.

Nyerere, Julius. 1968. *Ujamaa: Essays on socialism*. Dar es Salaam, Tanzania: Oxford University Press East Africa.

Ostrom, Elinor. 1990. *Governing the commons: The evolution of institutions for collective action*. Cambridge, UK: Cambridge University Press.

Parkipuny, Lazaro Ole. 1979. "Some crucial aspects of the Maasai predicament." In *African socialism in practice*, edited by Andrew Coulson. Nottingham, UK: Spokesman Books.

Pastoralists Nongovernmental Organisations Forum (PINGO). 2013. *Fact finding mission on the impact of wildlife investment in pastoralists areas of Monduli, Simanjiro, Babati, and Kondoa Districts from 5th June 2013 to 16th June 2013*. Arusha, Tanzania: PINGO.

Pittiglio, Claudia, Andrew K. Skidmore, Hein A. M. J. van Gils, and Herbert H. T. Prins. 2013. "Elephant response to spatial heterogeneity in a savanna landscape of northern Tanzania." *Ecography* 36(7): 819–831.

Prins, Herbert H. T. 1987. "Nature conservation as an integral part of optimal land use in East Africa: The case of the Masai ecosystem of northern Tanzania." *Biological Conservation* 40(2): 141–161.

Randilen Wildlife Management Area (RWMA). 2018. *Resource management zone plan 2018–2022*. Arusha, Tanzania: RWMA.

Raycraft, Justin. 2016. "Restrictions and resistance: An ethnographic study of marine park opposition in southeastern Tanzania." MA thesis. Department of Anthropology. University of British Columbia.

———. 2019. "Circumscribing communities: Marine conservation and territorialization in southeastern Tanzania." *Geoforum* 100: 128–143.

———. 2020. "The (un) making of marine park subjects: Environmentality and everyday resistance in a coastal Tanzanian village." *World Development* 126(104696): 1–12.

———. 2022a. "Community attitudes towards Randilen Wildlife Management Area." In *Tarangire: Human-wildlife coexistence in a fragmented ecosystem*, edited by Christian Kiffner, Derek Lee and Monica Bond, 109–128. New York, N.Y.: Springer.

———. 2022b. "Wildlife conservation through the lens of pastoralism: Institutional arrangements for rangeland management in the Maasai Steppe, Tanzania." PhD Dissertation. Department of Anthropology. McGill University.

———. 2023. "Wildlife and human safety in the Tarangire ecosystem, Tanzania." *Trees, Forests, and People* 13(100418): 1–11.

———. 2024a. "Perceived impacts of wildlife on agropastoral food production in northern Tanzania." *Ecology of Food & Nutrition* 63(3): 204–228.

———. 2024b. "Human-hyena (*Crocuta crocuta*) conflict in the Tarangire ecosystem, Tanzania." *Conservation* 4: 99–114.

Raycraft, Justin, George Tanner, and Edwin Maingo Ole. 2024. "Sharing landscapes with megaherbivores: Human-elephant interactions northeast of Tarangire National Park." *Environmental Challenges* (101005): 1–11.

Reid, Robin S. 2012. *Savannas of our birth: People, wildlife, and change in East Africa*. Berkeley, Calif.: University of California Press.

Rija, Alfan Abeid. 2009. "The long-term impacts of hunting on population viability of wild ungulates in Tarangire, northern Tanzania." MSc. thesis. School of Biological Sciences. Victoria University of Wellington.

Rodgers, Alan, Lota Melamari, and Fred Nelson. 2003. *Wildlife conservation in northern Tanzanian rangelands*. Symposium: Conservation in Crisis: Experiences and Prospects for Saving Africa's Natural Resources, Mweka College of African Wildlife Management, Tanzania, December 10–12, 2003. Arusha, Tanzania: TNRF.

Sachedina, Hassan. 2006. "Conservation, land rights and livelihoods in the Tarangire ecosystem of Tanzania: Increasing incentives for non-conservation compatible land use change through conservation policy." Pastoralism and Poverty Reduction in East Africa: A Policy Research Conference, Nairobi, Kenya, June 27–28, 2006.

———. 2008. "Wildlife is our oil: Conservation, livelihoods and NGOs in the Tarangire ecosystem, Tanzania." PhD Dissertation. School of Geography and the Environment. University of Oxford.

Sachedina, Hassan, and Fred Nelson. 2010. "Protected areas and community incentives in Savannah ecosystems: A case study of Tanzania's Maasai Steppe." *Oryx* 44(3): 390–398.

Sachedina, Hassan, and Pippa C. Trench. 2009. "Cattle and crops, tourism and tanzanite: Poverty, land-use change and conservation in Simanjiro District, Tanzania." In *Staying Maasai? Livelihood, conservation, and development in East African Rangelands*, edited by Katherine Homewood, Patti Kristjanson and Pippa C. Trench, 263–298. New York, N.Y.: Springer.

Scott, James C. 1985. *Weapons of the weak: Everyday forms of peasant resistance*. New Haven, Conn.: Yale University Press.

———. 1998. "Compulsory villagization in Tanzania: Aesthetics and miniaturization." In *Seeing like a state: How certain schemes to improve the human condition have failed*, 223–261. New Haven, Conn.: Yale University Press.

Shao, John. 1986. "The villagization program and the disruption of the ecological balance in Tanzania." *Canadian Journal of African Studies* 20(2): 219–239.

Shetler, Jan B. 2007. *Imagining Serengeti: A history of landscape memory in Tanzania from earliest times to the present*. Athens, Ohio: Ohio University Press.

Shivji, Issa G. 1986. *Law, state and the working class in Tanzania*. Portsmouth, N.H.: Heinemann.

Shivji, Issa G. 1998. *Not yet democracy: Reforming land tenure in Tanzania*. London, UK: IIED.

Shoo, Rehema Abeli, Elizabeth Kamili Mtui, Julius Modest Kimaro, Neema Robert Kinabo, Gladys Joseph Lendii, and Jafari R. Kideghesho. 2021. "Wildlife management areas in Tanzania: Vulnerability and survival amidst COVID-19." In *Managing wildlife in a changing world*, edited by Jafari R. Kideghesho, 1–17. London, UK: IntechOpen.

Singh, Neera M. 2013. "The affective labor of growing forests and the becoming of environmental subjects: Rethinking environmentality in Odisha, India." *Geoforum* 47: 189–198.

Spear, Thomas. 1997. *Mountain farmers: Moral economies of land & agricultural development in Arusha & Meru*. Berkeley, Calif.: University of California Press.

Spear, Thomas, and Derek Nurse. 1992. "Maasai farmers: The evolution of Arusha agriculture." *International Journal of African Historical Studies* 25(3): 481–503.

Spear, Thomas, and Richard Waller. 1993. *Being Maasai: Ethnicity and identity in East Africa*. Oxford, UK: James Currey.

Suich, Helen. 2010. "The livelihood impacts of the Namibian community based natural resource management programme: A meta-synthesis." *Environmental Conservation* 37(1): 45–53.

Sulle, Emmanuel. 2008. *Wildlife-based revenue transparency performance in Longido and Simanjiro Districts*. Arusha, Tanzania: Hakikazi Catalyst.

Sulle, Emmanuel, and Holti Banka. 2017. "Tourism taxation, politics and territorialisation in Tanzania's wildlife management." *Conservation & Society* 15(4): 465–473.

Sulle, Emmanuel, Edward Lekaita, and Fred Nelson. 2011. *From promise to performance? Wildlife management areas in northern Tanzania*. Arusha, Tanzania: TNRF.

Tanzania Natural Resource Forum (TNRF). 2008. *Meeting between wildlife division representatives and Simanjiro and Lolkisale stakeholders: Preliminary minutes*. Arusha, Tanzania: TNRF.

Trench, Pippa Chenevix, Steven Kiruswa, Fred Nelson, and Katherine Homewood. 2009. "Still 'People of Cattle'? Livelihoods, diversification and community conservation in Longido District." In *Staying Maasai?*, edited by Katherine Homewood, Patti Kristjanson and Pippa C. Trench, 217–256. New York, N.Y.: Springer.

Turner, Matthew. 1993. "Overstocking the range: A critical analysis of the environmental science of Sahelian pastoralism." *Economic Geography* 69(4): 402–421.

Vaccaro, Ismael, Oriol Beltran, and Pierre A. Paquet. 2013. "Political ecology and conservation policies: Some theoretical genealogies." *Journal of Political Ecology* 20(1): 255–272.

Watts, Michael. 2004. "Resource curse? Governmentality, oil and power in the Niger Delta, Nigeria." *Geopolitics* 9(1): 50–80.

Welch, Cameron. 2018. *"Land is life, conservancy is life": The San and the N‡a Jaqna Conservancy, Tsumkwe District West, Namibia*. Basel, Switzerland: Basler Afrika Bibliographien.

Weldemichel, Teklehaymanot G. 2020. "Othering pastoralists, state violence, and the remaking of boundaries in Tanzania's militarised wildlife conservation sector." *Antipode* 52(5): 1496–1518.

———. 2022. "Making land grabbable: Stealthy dispossessions by conservation in Ngorongoro Conservation Area, Tanzania." *Environment and Planning E: Nature and Space* 5(4): 2052–2072.

Wilfred, Paulo. 2019. "The challenges facing resident hunting in western Tanzania: The case of the Ugalla ecosystem." *European Journal of Wildlife Research* 65(6): 1–13.

Williams, Raymond. 1980. "Ideas of nature." In *Problems in materialism and culture: Selected essays*, 67–85. London, UK: Verso.

Wøien, Halvor, and Lewis Lama. 1999. *Market commerce as wildlife protector? Commercial initiatives in community conservation in Tanzania's northern rangelands*. London, UK: IIED.

Wolf, Eric R. 1982. *Europe and the people without history*. Berkeley, Calif.: University of California Press.

Woodhouse, Emily, and J. Terrence McCabe. 2018. "Well-being and conservation: Diversity and change in visions of a good life among the Maasai of northern Tanzania." *Ecology and Society* 23(1): 1–14.

Wright, V. Corey 2016. "Turbulent times: Fighting history today in Tanzania's trophy hunting spaces." *Journal of Contemporary African Studies* 34(1): 40–60.

———. 2017. "Turbulent terrains: The contradictions and politics of decentralised conservation." *Conservation & Society* 15(2): 157–167.

———. 2019. "Becoming enduimet & the precariousness of living with elephants." PhD Dissertation. Department of Anthropology. McGill University.

World Wildlife Fund (WWF). 2014. *Tanzania's wildlife management areas: A 2012 status report*. Dar es Salaam, Tanzania: WWF.

INDEX

AA. *See* Authorized Association
A104 Arusha-Babati highway, 59, 84, 97, 109
Acacia (plant genus), tree, 19, 156, 161
African socialism, forwarding vision of, 25
African Wildlife Foundation (AWF): community-based conservation initiatives, 7, 27; contracting Honeyguide, 109–10; discussions about WMA creation, 82–83; establishing WMA, 93–96; heartlands initiative, 73–75; politics of WMA formalization, 101–2; pressuring villages, 86–87; resource zoning plan, 114–15, 138; USAID contract, 108–9; WMA pilot period, 78
Agrawal, Arun, 14, 107, 132
Akie, people, 44
anti-poaching: activities, 108; efforts, 111–13; initiatives, 110–11, 129; model of conservation, 112–14; operations, 109, 124
Arusha (city): absentee herders and farmers living in, 76–77, 133; permit payments in, 160; safari company based in, 52; tanzanite market in, 101, 148–49
Arusha Declaration, 25. *See also* Arusha (city); Arusha (people)
Arusha (people), 16, 53, 58, 155, 158; commercial farmers and cattle barons, 99–102; complexities of community-based conservation, 155, 157, 160, 166–67, 169–72; conservation at crossroads, 173–83; encroachment of, 49–51; establishing WMA, 95–98; and foundations of social enterprise, 126–32; gratefulness of, 157; herding system trends, 31; historical patterns of in-migration, 31; involvement in WMA creation, 81–85; land issues, 46–47; and legislative reform, 73–77; new directions for WMA leadership and administration, 115, 118–20; politics of hunting and photographic tourism, 65–66; radical shift in attitudes of, 107–10; and ripple effect of WMAs, 77–80; sensitization in Mswakini and Naitolia, 87–93; surveying, 106–7; in Tarangire ecosystem, 16–20; and theory of change, 132–44
Authorized Association (AA), 178, 180–82; complexities of community-based conservation, 159, 162, 164–65; concerns of WMAs superseding tenure status of Maasai villages, 144–50; designing second resource management plan, 138–44; and local governance, 168; as main governance body of WMA, 117–24; new directions for WMA leadership and administration, 115; putting community first, 109–10; selecting members of, 135; and WMA establishment, 93–98; and WMA ripple effect, 78–79
AWF. *See* African Wildlife Foundation (AWF)

Banka, Holti, 153–54
Barabaig, people, 44
bed night fees, 59, 151–53, 159, 164; or head payment, 151, 153, 164
"being a good neighbor" (*kuwa ujirani mwema*), 48
Bell, Damian, 110–15, 131–32, 135, 137, 147, 161
"big five" (elephant, rhino, lion, buffalo, and leopard), 32–33
Big Life Foundation, 110–11, 114
Board of Trustees, 117, 121
bomas (pastoral homesteads), 21, 34, 37, 97, 101, 143, 145–46

203

Index

Borner, Markus, 47, 50, 64
Boundary Hill: gate, 162–63; investment sites near, 165; poaching near, 69
Boundary Hill Lodge: controversies with, 78; court case, 90–91; and distrust, 57; economic complexity, 156; end of drawn-out conflict with, 156–57; and entrance fees, 152–53; exclusion of, 70; interviewing lodge manager of, 158; negotiating village-based photographic tourism concession with, 66–67; raising funds for, 53–54; sentiments developing despite shortcomings of, 63; struggles of, 56–57; waiting for construction of, 58–59. *See also* Treetops Lodge
Britain, building on German groundwork, 24
broken pot curse, 84–85
buffer zone, 36, 49, 55, 70–71
Bundu Safaris Ltd., 64–66, 70–71, 80
Burunge WMA, 74, 92, 113–15, 137, 140, 164

CAMPFIRE. *See* Communal Areas Management Programme for Indigenous Resources
capital, wildlife as: joint ventures between tour operators and villages, 4; key governance challenge, 33; neoliberal conservation struggles, 34–36; new value attribution, 33; outcomes of the direct investment model, 35; pastoral areas adjacent to national parks, 33–34; privatization of conservation practice, 34–35; safari tourism complicating political dynamics of Maasailand, 32–33
cattle barons, 99–102, 133
CCROs. *See* Certificates of Customary Rights of Occupancy
Certificates of Customary Rights of Occupancy (CCROs): commercial farmers and cattle barons, 99–100; communal CCROs, 31–32; confidence in, 42; enforcement of, 32; obtaining, 7–8; representation of villages, 30–31; securing, 41; struggle for sovereignty, 30–32; and wildlife, 32, 124–25
Chama Cha Mapinduzi, political party, 89–90
charcoal: burning, 139, 143; production of, 47, 67, 98, 143
Chem Chem Lodge, 137
Cochran, William G., 22
College of African Wildlife Management Mweka, 74, 115

commercial farmers, 176; and absentee landholders, 99–102; competing interests from, 76–77; crackdowns on, 133–35
Commiphora (plant genus), bushland, 19
Commissioner of Land, 31
Communal Areas Management Programme for Indigenous Resources (CAMPFIRE), 5–6
community: and anti-poaching initiatives, 112–15; and arrival of Honeyguide, 110–12; attitudes toward Randilen WMA, 103–7; communication between WMA and, 131–32; key to unlocking support from, 126–32; policing, 96–97; putting first, 107–15; reasons for supporting WMA, 107–8; reducing crop damage for, 127–31; shepherding elephants, 130–31; shifting conservation sentiments, 108–10; and theory of change, 132–37. *See also various entries*
community-based conservation: and bed night fees, 151–52; Boundary Hill Lodge Co. Ltd. as form of, 53–54; communication as important part of, 131–32; dealing with corruption, 165–70; economic viability of WMA model, 160–65; ending Boundary Hill Lodge dispute, 156–57; enhancing local capacity for, 116; environmental protection model as form of, 141; environmentality in context of, 14; findings running contrary to notion of, 153–54; and formalization, 77; further economic complexity in, 151–52, 154, 156; institutions bearing influence on, 12; introduction of entrance fees, 152–53; and LCA, 55–56; local support for, 63–64; lodge/camp security, 160; model of, 19–20, 73, 110–13; planting seedlings of, 34; pre-WMA era of community-based conservation, 151–52; Randilen WMA as, 104, 121, 179, 183; recentralization, 39, 86; recognizing centrality of tourism, 157–60; refining vision of, 111; scholarly critique of, 174–75, 178; semi-community-based conservation area, 109; socioeconomic effects, 170–72; struggle for sovereignty, 42; tourism revenue distribution across numerous villages, 154–55; toward, 5–10; WMA benefit distribution, 155–56; WMA-level fees, 152–53
community-based organization (CBO), 40–41, 93–94, 96, 109, 115
concessions: economics of, 164; for hunting, 72; overlaps with GCA, 66–67, 70–72, 78;

payments, 151; for photographic tourism, 5, 29, 34, 35, 37, 52–54; 56, 58–59, 75, 110
conservation: community-based conservation, 151–72; in context, 1–4; at crossroads, 173–83; foundations of social enterprise, 126–50; growing pains in field of, 2; having conservation in common, 9–12; Lolkisale land squeeze, 44–63; Maasai society and state, 24–43; Maasailand and, 16–20; rise of Randilen, 103–25; struggle for sovereignty, 28–32; toward community-based conservation, 5–9; trophy hunters and photographic tour operators, 64–80; WMA creation, 81–102
conservation in common, having, 9–12, 128, 174
contestations, Lolkisale GCA/LCA: buffer zones, 70; consensus-based solution, 71; enforcing buffer zones, 70–71; institutional scramble, 72; meetings for resolving, 69–70; safety proposals, 70; wildlife activity fee, 71–72
contributions (*michango*), 170–72
corruption: examining contrasting cases, 168–69; good governance of Lolkisale village regarding, 165–66; local governance in Nafco, 165–68; repercussions of poor village leadership, 169–70; resident hunting and, 68; risk of, 123
crops, reducing damage to, 127–31
crossroads, conservation at: cultural politics of transformation, 177–79; formalization, 179–82; key takeaways, 182–83; overview, 173–76; pastoral resilience, 179–82

decentralization, politics of, 36–39
de Sardan, Jean-Pierre Olivier, 15
Descartes, René, 2
de Soto, Hernando, 33
dispersals (of wildlife), managing, 47–51
District Advisory Board, 117, 121
District Authority Review, 31
drought reserve, 141–42, 180

East African Safari & Touring Company (EASTCO), 52–53, 56, 57
EASTCO. *See* East African Safari & Touring Company
economic viability, WMA model: achieving financial independence, 161–62; codification, 160–61; constraint awareness, 163–64; investor quality, 164–65; lack of harmonization, 162–63

ecosystem services, payment for, 34
elephants, 5, 33, 65, 125, 135, 160, 178; crop-raiding, 11, 113, 130–31, 169, 179; herding back into WMA, 91; human-elephant conflict (HEC), 127–29; importance of, 61–63; important dispersal area for, 9, 55–56, 78, 81, 142; managing dispersal of wildlife, 49–50; milk of, 60–61, 63, 80, 154, 183; in northern Tanzania, 47; poaching, 109–10, 130; and putting community first, 121–22, 129–31; and secure tenure rights, 127; shepherding, 130–31; subpopulation of, 19, 56
Emboreet (village), 66
Empathetic conservation: conceptualization, 174; Honeyguide's efforts as example of, 133, 137; and local values, 10, 128; unlocking community support, 126
Enduimet WMA, 41, 73–74, 86; anti-poaching track record in, 109, 111; and ripple effect of WMAs, 77–80
entrance fees, introduction of, 152–53
environment, theorizing as political arena, 13–14
environmentality, concept, 14; and rise of community support, 107, 132, 173, 174
environmental governance, concept, 15
ethnographic field research, methodology for investigating, 20–23

fieldwork, 15, 58, 88, 103, 107, 110, 119, 128–30, 134, 143, 160–64, 168, 175
First World War, 24
Fletcher, Robert, 14
Forestry Policy of Tanzania, 66
formalization: of bed night fees, 153–54, 178; and community-based conservation, 8, 11; of game reserves, 24; and pastoral resilience, 179–82; of Randilen WMA, 76, 80, 89–90, 93, 124, 153, 179–82; and security, 160; of villages, 30
fortress conservation, 3–7, 9, 13, 24, 33, 104, 141, 149, 175, 177–78
fortress model of conservation, 22, 116, 174
Foucault, Michel, 14, 133, 174

Game Controlled Areas (GCAs), 5, 64–65, 68, 175; classification, 64; complications of, 36–37; converting into game reserves, 42; difference from game reserves, 65; and legislative reform, 7; patrolling, 65; and politics of decentralization, 36–38; and social implications of villagization, 26–27

Game Ordinances, 24
game reserve, 7, 8, 24, 26, 33, 36–38, 42, 65, 68, 125, 175, 177
Gardner, Benjamin, 14, 27, 35, 177
GCAs. *See* Game Controlled Areas
GEF. *See* Global Environment Facility
German Imperial Decree of November 26, 1895, 24
Global Environment Facility (GEF), 54, 56, 66
governance, building local capacity for: benefits of formalization, 124–25; committee synergy, 119–21; developing AA into strong institution, 118–20; facilitating series of trainings, 118; general AA meetings, 121–23; main governance body, 117; protection activities, 123–24; village council members, 117–18
governmentality, concept, 14

Halcyon Tanzania Ltd., 58
Hardin, Garret, 4
"heartland" initiative, 73–74
HEC. *See* human-elephant conflict
Hodgson, Dorothy L., 14
Honeyguide, 11, 108, 178–79; anti-poaching track record of, 109–10; building local capacity for governance and management, 117–25; and economic viability of WMA model, 160–64; fostering management strategy, 126–27; new directions for WMA leadership, 115–16; programmatic aims of, 109; putting community first, 108–15; reducing crop damages, 127–31, 169; and revised management plan, 138; and theory of change, 132–37
human-elephant conflict (HEC), 127–29
hunting, politics of: contestations over Lolkisale GCA, 69–72; legislative reform, 72–77; overview, 64–69; ripple effect of WMAs in Maasailand, 77–80

ilaigwenak (Maasai traditional leaders), 77
Indigenous: communities, 5–6, 174; hunters, 111; land claims, 177, 187; peoples, 1–2; rights, 82; stewardship, 26
institutional bricolage, 15
institutions, theme, 12–16. *See also various entries*
Integrated Conservation Management Plan, 54–55
International Monetary Fund, 27

kawaida (neutral category). *See* community, attitudes toward Randilen WMA
Keane, Aidan, 172
Kilimanjaro Heartland, 73
KIPOC. *See* Korongoro Integrated People's Oriented to Conservation
Kisongo Maasai, people: and Arusha interactions, 18–19; commercial farmers and cattle barons, 99–100; complexities of community-based conservation and, 151–72; concerns about WMAs superseding tenure status, 144–49; conservation at crossroads, 173–83; and foundations of social enterprise, 126–31; inhabiting Randilen WMA, 19–20; migrating to territories of, 31; and new tourism frontiers, 51–63; Operation *imparnati* ("permanent habitation" or sedentarization) affecting, 46; as "people of cattle," 16; and politics of hunting and photographic tourism, 75–77; prior to colonialism, 44; resenting Arusha encroachment, 51; and ripple effect of WMAs in Maasailand, 77–80; sharing institutions with, 18; and theory of change, 133–34; views on new management plan, 138, 142–44; WMA creation and, 43, 81–86, 90, 95
Korongoro Integrated Peoples Oriented to Conservation (KIPOC), 28

Lake Manyara National Park (Lake Manyara NP), 19, 51
land, theme, 12–16
Land Act No. 4 of 1999, 29, 66
land squeeze. *See* Lolkisale land squeeze
LCA. *See* Lolkisale Conservation Area
leadership, WMA, 115–16
legislative reform: AWF fashioning regional program, 73–74; buffer zone enforcement, 71; and commercial farmers, 76–77; perceiving WMAs with distrust, 28, 75; revised WCA No. 5 of 2009, 72–73; weighing risks and benefits of establishing WMA, 75–76; well-intentioned work, 74–75; WMA framework, 151, 165–66
Lemooti (village), 21, 45, 53, 61, 76, 81, 83, 95–96, 99, 101, 177; building trust in, 133–34; community attitudes toward Randilen WMA, 106–7, 140; elephants in, 129; and revised management plan, 138, 142–44; *tajiri mwenye kiti* ("rich village chair") in, 144–50

Lengoolwa (village), 21, 53, 77, 81–84, 86, 116, 118–19, 127, 138, 146–47, 149, 154, 170
Lesorogol, Carolyn, 60–61
livestock: concerns about WMAs superseding village tenure status, 144–50; as first priority, 150; formalizing grazing system, 138–39; grazing area, 98, 107, 124, 171; keepers of, 85, 97, 100, 126–27, 138, 140–42, 144, 146, 171, 180–81; legislation related to, 72–77; and management of wildlife dispersal, 47–51; in mixed-use zones, 142–44; new grazing plan for, 139–41; and new tourism frontier, 51–52, 55; as secondary consideration, 82
LLWZ. *See* Lolkisale Livestock and Wildlife Zone
lodges: and complexities of community-based conservation, 152–65; and Lolkisale land squeeze, 52–59. *See also* Boundary Hill Lodge; Treetops Lodge
Loibor Siret (village), 66
Loiborsoit (village), 58, 66
Loliondo: division, 7, 8; comparisons with Longido, 78; conflicts with Tanzania Breweries in, 29; fears about WMAs, 7, 146–47; GCA, 37, 175; Maasai of, 124; photographic wildlife tourism in, 35, 109; securing CCROs, 8, 41–43
Lolkisale (village), 90; fieldwork in, 60. *See also* Lolkisale Conservation Area (LCA)
Lolkisale Conservation Area (LCA), 52, 139; Boundary Hill Lodge struggles, 56–58; conserving on own accord, 63; contestations over, 69–72; establishing joint conservation venture, 52–53; frictions emerging at Naitolia Camp, 58; impetus behind, 52–55; interest in community lands for tourism, 54; initial proposal of, 66–67; key aspect of expanding, 55–56; lobbying for inclusion of elephant habitat area in, 55–56; original proposal of, 66–67; and public buses, 59–61; raising funds for building lodge, 53–54; reinvesting bed night fees into management of, 59; revenue from Treetops, 59; scaling up, 54; success as "win-win" joint venture, 58–59; weighing benefits and risks of, 55; zones protected by, 56
Lolkisale GCA, 50, 55, 65–66; contestations over, 69–72; discontinuities in formal status of land in, 67; superseding, 72–73; using to grab wildlife revenue, 68
Lolkisale land squeeze: managing dispersing wildlife, 47–51; new tourism frontier, 51–63; overview, 44–47
Lolkisale Livestock and Wildlife Zone (LLWZ), 55–56, 76–77
Lolkisale Village Council (LVC), 52–53, 58, 60, 63
Longido (district): Enduimet WMA in, 175; heartland initiative, 73; joining Enduimet WMA in, 8, 41–43; Kilimanjaro, NGO collaborations in, 110; subdividing Maasailand, 16; visiting Kisongo of, 77–79, 86
Longido GCA, 78
Loveless, Stephanie, 87–88; 91–92; 98–102; 106–7; 117, 122, 134
Lowassa, Edward, 89–92; 94; 102
LVC. *See* Lolkisale Village Council

Maasai (people), ethnographic context, 16–20. *See also various entries*
Maasailand, 16–20; community-based conservation in, 151–72; creating WMA in, 81–102; foundations of social enterprise, 126–50; historical context, 24–27; Lolkisale land squeeze, 44–63; Maasai society and state, 24–43; ripple effect of WMAs in, 77–80; rise of Randilen in, 103–25; struggle for sovereignty, 28–32; trophy hunters and photographic tour operators, 64–80; wildlife as capital in, 32–36; and WMAs, 39–43; *See also* conservation
Maasai Member of Parliament, 29
Maasai society and state: flexibility of presenting as "Maasai," 31; overview, 24–27; politics of decentralization, 36–39; recentralization and WMAs, 39–43; struggle for sovereignty, 28–32; wildlife as capital, 32–36
Maasai Steppe, 20, 47, 73–74. *See also various entries*
Maasai Steppe Heartland, 73
maendeleo ("development"), 48, 59, 121, 148
Magufuli, John, 37, 89, 104
Makuyuni (village), 55–56, 119, 136. *See also* elephants; Makuyuni Elephant Dispersal Area (MEDA)
Makuyuni Elephant Dispersal Area (MEDA), 55–56; additional tourist camp proposed in, 58; and *ndoroboni* (communal grazing area), 85; WMA including, 86. *See also* elephants
Makuyuni River, 58
Makuyuni Wildlife Park, 95–96

management, building local capacity for: committee synergy, 119–21; developing AA into strong institution, 118–20; facilitating series of trainings, 118; general association meetings, 121–23; main governance body, 117; protection activities, 123–24; village council members, 117–18
management, concept, 14–15; examples of, 15–16
Manyara Ranch, 109, 129
Masai Reserve, creation of, 24, 51
Mawanja, Seraphino, 90–91, 102
maziwa ya tembo ("milk of elephants"), 60–61, 63, 80, 154, 183
Mbugwe, people, 44
McCabe, J. Terrence, 141–42
MDC. *See* Monduli District Council
MEDA. *See* Makuyuni Elephant Dispersal Area
Melembuki, Meshurie, 115–16
member villages. *See* villages
Mererani, tanzanite mines in, 100–101, 148
michango. See contributions
milk of elephants. See *maziwa ya tembo*
Ministry of Livestock, 50
Ministry of Natural Resources and Tourism (MNRT), 6–7, 64, 86; commitment to WMA model, 86–87; lack of harmonization, 162–63; letter of endorsement from, 66–67; and politics of decentralization, 36–39, 69
mixed-use zones, 142–44, 146
MNRT. *See* Ministry of Natural Resources and Tourism
MODECO. *See* Monduli Development Corporation
Mollel, Samwel Saruni, 116
Monduli Development Corporation (MODECO), 46
Monduli District: and AWF, 87; Government, 82–83, 92, 94–95, 152; land dynamics and resource use in rural, 99, 143, 144; Lolkisale GCA in, 65–66; MP of, 89; Randilen WMA in, 19; rangeland fragmentation in, 31; subdividing Maasailand, 16. *See also* Monduli District Council
Monduli District Council (MDC), 20, 64, 67–70, 154, 161
Monduli District Game Officer (Monduli DGO), 67, 90
Mount Meru, 19
MSP. *See* Multiple Stakeholder Partnership
Mswakini (villages), 21, 47–48, 54, 58; AA members from, 117–19; commercial farmers and cattle barons, 99–102; establishing WMA, 81–83, 86–89, 91–99, 101–2; fieldwork in, 103–7; and foundations of social enterprise, 127, 129–30, 133–36, 142–43; putting community first in, 107–15; ripple effects of WMAs in Maasailand, 77; sensitizing, 87–93
Mswakini Chini (village), 21, 77, 92, 96, 141, 144; community attitudes toward Randilen WMA, 106–7; and complexities of community-based conservation, 154, 156–57; and misinformation, 138–39; and WMA creation, 81–83, 86–87
Mswakini Juu (village), 21, 48, 54, 58, 96, 98, 115, 129, 136, 139, 171; community attitudes toward Randilen WMA, 106–7; and complexities of community-based conservation, 154–56; sensitization of, 91–93; and WMA creation, 81–83, 86–87
Mto wa Mbu (town), 42, 90, 121, 122, 147
Multiple Stakeholder Partnership (MSP), 54
Mwinyi, Ali Hassan, 27

NAFCO. *See* National Agricultural and Food Corporation
Nafco (village), 46, 90–91, 130–31; corruption in, 165–70
Naitolia (village), 21, 46, 154–56, 171; AA members from, 117–19; buffer zones around, 70; commercial farmers and cattle barons, 99–102; establishing WMA, 93–98; fieldwork in, 103–7; and foundations of social enterprise, 127, 129–30, 133–36, 142–43; Naitolia Concession, 54, 56, 58; and new tourism frontier, 51–63; putting community first in, 107–15; sensitizing, 87–93; and WMA creation, 81–87
Naitolia Camp, 54, 57–58, 70, 81, 93, 156
Naitolia Concession, 54, 56, 58
Namibia, 5
National Agricultural and Food Corporation (NAFCO), 46–47
National Park Ordinances, 24
national parks, 29, 49, 55, 159, 173–75, 179; anti-poaching model of conservation taking inspiration from, 112–13; becoming synonymous with conservation, 2; expanding, 24; and politics of decentralization, 36–39; and politics of hunting and photographic tourism, 64, 73–75; revenue outside, 7; and

struggle for sovereignty, 32; in Tanzania, 5; wildlife as capital in, 32–33; and WMA creation, 90, 98. *See also various entries*
National Villages and Ujamaa Villages Act of 1975, 25
National Villages Land Act, 29, 30, 37
Nature Conservancy, the, 114, 162, 164
NCA. *See* Ngorongoro Conservation Area
NCAA. *See* Ngorongoro Conservation Area Authority
Ndoroboni (communal grazing area in Makuyuni village), 55, 85
neighborhood villagization. *See ujamaa vijijini*
neoliberal conservation, 27, 34–35, 39
neoliberalization, contradicting principles of, 6
neutral attitude, category. *See* community, attitude toward Randilen WMA
Ngorongoro Conservation Area Authority (NCAA), 36–39, 42
Ngorongoro Conservation Area (NCA), 24, 26, 38, 50, 92, 134
NGOs. *See* nongovernmental organizations
nongovernmental organizations (NGOs), 7, 28, 41–42, 50, 61, 99, 108–18, 123, 130, 132, 137, 157, 163, 173
Northern Hunting, 78
Northern Tanzania Rangelands Initiative, 163
Nyerere, Julius, 25–27, 38

OBC. *See* Ortello Business Corporation
Oldonyo (village), 21, 53, 81, 83, 106, 119, 126–27, 146–47, 154, 165–66, 170
"open access" property regimes, 4
open areas, 26–27, 36, 48, 65, 68–69
Operation *imparnati* ("permanent habitation"), 25, 46
Ortello Business Corporation (OBC), 8, 37

Parkipuny, Lazaro Ole, 28–29
Parliament of Tanzania, 46
pastoralists, 4, 7, 55, 85, 173, 177; building local capacity for governance/management, 117, 124, 126–27; choosing from villages, 140; concerns with TANAPA, 48–51; constraining, 44; defending, 90; local, 48, 55, 90, 124, 141, 180, 182; in Maasailand, 16–20; modernizing, 25; opening photographic zone to, 141; overlapping jurisdictions, 144–50; and politics of decentralization, 36, 38; putting community first, 107; resilience, 179–82; serving economic interests of, 155; and struggle for sovereignty, 28–32; and three central themes, 12–13; underlying question for, 8; and wildlife as capital, 32, 35
pastoral resilience, 179–82
people of cattle, 16, 126. *See also* Kisongo Maasai, people
photographic tourism, 5, 8, 37–38, 53, 181; concession negotiation, 66–67; establishing first village-based tourism ventures, 109; legislative reform, 72–77; Lolkisale GCA/LCA, 69–72; overview, 64–69; permissible investments, 94; prohibiting, 67; reclassifying areas for, 90; resident hunting conflicting with, 68–69; ripple effect of WMAs in Maasailand, 77–80; setting aside acres for, 54. *See also* tourism
poaching. *See* anti-poaching
Pololeti Game Reserve, 7–8, 37, 42, 175
public buses, struggles of, 59–61

Randilen WMA: building local capacity for governance and management, 117–25; central themes of, 12–16; community attitudes toward, 104–7; concerns about WMAs superseding village tenure status, 144–50; conflicting cases of corruption, 165–70; and conservation at crossroads, 173–74; cultural politics of transformation, 177–79; economic viability, 160–65; formalization as key feature of, 179–82; inhabitants of, 16–18; key takeaways, 182–83; location of, 19–21; new directions for leadership and administration, 115–16; overview, 9–12; providing "quantitative snapshot" of community sentiment toward, 103–7; putting community fist, 107–15; recognizing centrality of tourism in, 157–60; scholarly literature on, 22–23; socioeconomic effects, 170–72; and theory of change, 132–37
rangelands, managing, 13, 26, 140–41, 144, 146, 149, 173, 179
range sciences, broadening of, 4
recentralization, and WMAs, 39–43
Regulation of Land Tenure Act of 1992, 66
research fatigue, 104
resident hunting, 68–69
resilience, concept, 179–82
resource zoning management plan (RZMP), 93–96

Rift Valley Seed Company Ltd., 44–47
rights claims, new formal framework for articulating, 30
Rodgers, William A., 63
rules of the game, term, 12
RZMP. *See* resource zoning management plan

Sachedina, Hassan, 66, 75
Sanford, George (pseudonym), 74–75
savannas, 3, 16, 33
Scott, James C., 26, 40
second resource management plan (2018–2023): correcting misinformation, 138–39; drought reserve, 140–42; establishing two primary WMA zones, 139–41; mixed-use zone classification, 142–44
selective adoption, concept, 15
sensitization: and Arusha-led village councils, 92–93; cookie-cutter responses, 87–88; lack of awareness of WMA, 88–89; major politicians involved in, 89–91; money as incentive, 91–92; purpose of, 87
Serengeti National Park, 24
Shu'mata Camp, 78–79
Sidai Camp, 53, 156
sidetracking, concept, 15
Simanjiro District (Manyara Region), 48, 66, 100
Sluis Brothers Ltd., 46
social enterprise, foundations of: communication between WMA and community, 131–32; concerns about WMAs superseding village tenure status, 144–50; designing second resource management plan, 138–44; fostering empathetic management strategy, 126–27; reducing crop damage, 127–31; securing donor funds, 129–30; shepherding elephants, 130–31; theory of change, 132–37
socioeconomic effects, WMA model, 170–72
Sokwe Camps, 109–10
sovereignty, struggle for: alienation of pastoral lands, 28; challenge of using CCROs, 41; conflict in Loliondo between villages and state, 29–30; new formal framework for articulating rights claims, 30; pitfalls of individualized land tenure, 31; Tanzania Breweries Limited example, 29; villages working within constraints of formal policy, 28; WMAs against backdrop of, 43
Spear, Thomas, 19

Steyn, Hermanus, 44, 46, 64–65. *See also* Rift Valley Seed Company Ltd.
stock seeds, 46
sub-Saharan Africa, rise of protected areas in, 2
Sulle, Emmanuel, 63, 153–54

tajiri mwenye kiti ("rich village chair"), 145–48, 150
TANAPA. *See* Tanzania National Parks Authority
Tanganyika, 24. *See also* Tanzania
Tanzania: classifying community lands in, 5–6; community-based conservation in, 151–72; consequences of structural adjustment program in, 27; conservation at crossroads in, 173–83; conservation becoming self-fulfilling prophecy in, 3–4; creating WMA in, 81–102; emergence of protected areas in, 2; foundations of social enterprise, 126–50; having conservation in common in, 9–12; introducing WMAs in, 6–7; Lolkisale land squeeze, 44–63; Maasailand in, 16–20; Maasai society and state, 24–43; politics of decentralization in, 36–39; rise of Randilen in, 103–25; struggle for sovereignty, 28–32; studying conservation in, 2–3; toward community-based conservation, 5–9; trophy hunters and photographic tour operators, 64–80. *See also* conservation; Wildlife Management Areas (WMAs)
Tanzania Breweries Limited, 29
Tanzania National Parks Authority (TANAPA): capturing emergent market, 52; extracting value from wildlife, 64–65, 72; lack of policy harmonization, 162–63; managing dispersal of wildlife, 47–51; and politics of decentralization, 36–39, 42, 162–63; services provided by, 116
Tanzania Natural Resources Forum, 70
Tanzania Revenue Authority (TRA): and lodges, 159; and taxation, 151, 152
Tanzania Wildlife Management Authority (TAWA): collecting revenue, 124, 134, 178; and community-based conservation, 154, 159, 161, 163; formalized tax, 152; income-sharing between villages and, 83; and politics of decentralization, 36–39; politics of hunting and photographic tourism, 67–69, 71–72, 75–76, 78; and recentralization, 39–42; reductions in revenue percentage kept by, 176; resent-

ment toward, 68; resident hunting conflicting with tourism activities, 68–69; role in WMA creation, 83, 85, 90, 94, 96
Tanzanite: extracting, 100–101; market, 148; trade, 145
Tarangire Conservation Area, 56–58, 81, 93–94, 156
Tarangire Conservation Area Management Plan, 58
Tarangire Conservation Co. Ltd. (TCCL), 53, 56
Tarangire Conservation Project, 50
Tarangire ecosystem: fragmentation of, 63; human-wildlife conflict in, 49; and Maasailand, 16–21; and Maasai Steppe Heartland, 73, 74; tourism experiences in, 52; wildlife dispersals in, 33, 51–52
Tarangire Elephant Project, 55–56
Tarangire National Park (Tarangire NP), 9, 19, 31, 78, 129, 139, 156, 162–63; conflicts over Lolkisale GCA/LCA, 69–72; and legislative reform, 74–77; and Lolkisale land squeeze, 44–47; managing wildlife dispersals, 47–51; new tourism frontier, 51–55, 63; politics of hunting and photographic tourism, 64–66; and Randilen WMA, 19–20, 179, 182; wildlife congregating in, 20
TAWA. *See* Tanzania Wildlife Management Authority
TCCL. *See* Tarangire Conservation Co. Ltd.
tenure status, concerns about superseding, 144–50
theory of change, Honeyguide's: building up trust of people, 133–35; overview, 132–33; targeting village chairs, 135–37; three main goals, 137
tourism: arrangements, 52, 75; camps, 70, 118; concessions, 66, 75, 78; infrastructure, 160; investments in, 6, 27, 42, 75, 94; management plans, 110; new frontier of, 51–63; new selling points for, 51–52; photographic, 5, 8, 37–38, 53–54, 66–67, 74–75, 78, 82, 90, 94, 181; politics of, 64–80; rapid growth of, 51–52; recognizing centrality of, 157–60; revenues from, 6, 53, 57–59, 113, 154, 156, 161, 163; sector, 2, 158; unit, 123–24; wildlife, 32–36
Tourism and Photographic Management Zone (TPMZ), 139
TPMZ. *See* Tourism and Photographic Management Zone
TRA. *See* Tanzania Revenue Authority
tragedy of the commons, theory, 4

transformation, cultural politics of, 177–79
Treetops Lodge: creating economic complexity, 156; dispute between Bundu Safaris and, 70–72; foundations of social enterprise, 129; and GCA classification, 91; heavy lifting in terms of revenue, 159, 164–65; Lolkisale arrangement with, 75–76, 80; and new tourism frontier, 58–60, 63; paying bed night fees, 151, 153–54; and politics of hunting and photographic tourism, 65–68; putting at risk private contract with, 80–82; and Revised WCA of 2009, 72–73; tapping into agreement with, 85–86; and WMA creation, 80–82, 85–86
trophy hunting: concession negotiation, 66–67; legislative reform, 72–77; Lolkisale GCA/LCA, 69–72; low-quality game, 68; managing block for: 8, 26, 36–38, 67, 90, 152; overview, 64–69; representing big money, 65; resident hunting conflicting with, 68–69; ripple effect of WMAs in Maasailand, 77–80
trust, building up, 133–35

UCRT. *See* Ujamaa Community Resource Team
Ujamaa ("extended family" or African socialism), 25
Ujamaa Community Resource Team (UCRT): collaborating with Honeyguide, 114; concerns about WMAs, 41–42; protecting *ndoroboni*, 85; securing CCROs, 7, 30–31
ujamaa vijijini ("neighborhood villagization"), 25

Vachellia (plant genus), 19
vehicles, struggles of, 59–61
VGS. *See* village game scouts
village game scouts (VGS), 115, 133, 135, 150, 152; carrying out protection activities, 123–25; and complexities of community-based conservation, 156–58; interviewing, 130–31; and revised Management Zone Plan, 139–44; salaries of, 162–63
Village Land Act, 29, 38, 66, 72, 79, 99, 146
villages, 81–82, 86–87; classifying community land as, 5–6; conflict in Loliondo between state and, 29–30; formalization of, 30–31; lamenting empty promises, 48; and National Villages and Ujamaa Villages Act of 1975, 25; politics of decentralization, 36–39; subvillages, 19, 22, 53, 79, 83–84, 103–4, 145, 153; and wildlife as capital, 32–36; working within constraint of formal policy, 28

villagist, term, 147
villagization: and Arusha encroachment, 31, 46, 65; disregarding seasonality and cultural adaptations, 26; disrupting stability, 26; and formalization, 5, 180; as initiative, 25; social implications of, 26–27; legacy of, 28, 29, 59, 67
vitongoji ("sub-villages"), 21

walkabouts, 21
Ward Development Council, 55
WCA. *See* Wildlife Conservation Act No. 5 of 2009; Wildlife Conservation Act of 1974; Wildlife Conservation Act of 2009
wilderness: hegemonic influence of, 3; experiences, 52; and fortress conservation, 13, 33, 44, 128
wildlife: activity fee, 71–72; as capital, 32–36; conservation, 5, 12–13, 20, 24, 41, 48, 68, 73–74, 77, 92, 110, 129, 151, 175, 179–80, 182; human-wildlife conflict, 36, 49, 109, 111, 113, 178; human-wildlife interactions, 178; managing dispersal of, 47–51; policies, 32, 41–42, 67; and politics of decentralization, 36–39; populations, 32–33, 68, 70, 80, 90, 172; resources, 6, 38, 40, 65, 69–72, 74, 173, 175; ripple effect of WMAs in Maasailand, 77–80; as state property, 72; as theme, 12–16; tourism, 6, 8, 11, 32, 34–36, 38–39, 41–42. *See also* Wildlife Management Areas (WMA)
Wildlife Conservation Act No. 5 of 2009, 118
Wildlife Conservation Act of 1974, 26, 37, 64
Wildlife Conservation Act of 2009, 86, 118, 153
Wildlife Conservation Regulations, 117
Wildlife Conservation Society, 55
Wildlife Management Areas (WMAs), 103, 126, 173; accepting model of, 8; arguments about, 9; backdrop for introducing, 36–39; boundary disputes, 95–96; building up trust in, 133–35; cattle barons, 99–102; central government as reason for reforming LCA into, 86–87; codification of, 124–25; commercial farmers, 99–102; communication between community and, 131–32; concerns about WMAs superseding village tenure status, 144–50; and conservation at crossroads, 173–83; contrasting cases of corruption, 165–70; and direct investments, 42; economic viability of, 160–65; establishing, 93–98; existential dilemma of pastoralists, 42–43; formalization of, 8; income-sharing between villages, 83; intervillage politics and, 81–82; introduction of, 6–7; invisible hand behind concept of, 73; as microcosms of wider sectoral reform, 40–41; moving forward with, 84; new directions for leadership and administration, 115–16; perceiving with distrust, 75; potential and risks of, 41–42; potential for villages, 82–83; preparing resource zoning management plan for, 93–95; procedures for establishing, 41; reasons for not joining, 84–86; recentralization and, 39–43; recognizing centrality of tourism in, 157–60; reconsolidation of control over wildlife tourism, 39–40; ripple effect of, 77–80; rising tensions, 96–97; and sensitization, 87–93; socioeconomic effects, 170–72; and theory of change, 132–37; top-down enforcement of zoning scheme of, 97–98; village chair arrests, 98; ward-level politics and, 85–86; weighing risks and benefits of establishing, 75–76; withdrawing from, 86
Wildlife Policy of 1998, 6, 40–42, 66–67, 73, 75
Wilfred, Paulo, 69
WMA Financial Viability Tool, 161–62
WMA Regulations, 118
WMAs. *See* Wildlife Management Areas
World Bank, 27, 56; and International Finance Corporation, 53–54; loan transfer to Lolkisale village, 57–58
World Wildlife Fund (WWF), 42, 73–74, 109, 114
Wright, V. Corey, 8, 41, 77–78, 175
WWF. *See* World Wildlife Fund

Yellowstone National Park, 2

Zimbabwe, 5–6

GEOGRAPHIES OF JUSTICE AND SOCIAL TRANSFORMATION

1. *Social Justice and the City, rev. ed.*
   BY DAVID HARVEY
2. *Begging as a Path to Progress: Indigenous Women and Children and the Struggle for Ecuador's Urban Spaces*
   BY KATE SWANSON
3. *Making the San Fernando Valley: Rural Landscapes, Urban Development, and White Privilege*
   BY LAURA R. BARRACLOUGH
4. *Company Towns in the Americas: Landscape, Power, and Working-Class Communities*
   EDITED BY OLIVER J. DINIUS AND ANGELA VERGARA
5. *Tremé: Race and Place in a New Orleans Neighborhood*
   BY MICHAEL E. CRUTCHER JR.
6. *Bloomberg's New York: Class and Governance in the Luxury City*
   BY JULIAN BRASH
7. *Roppongi Crossing: The Demise of a Tokyo Nightclub District and the Reshaping of a Global City*
   BY ROMAN ADRIAN CYBRIWSKY
8. *Fitzgerald: Geography of a Revolution*
   BY WILLIAM BUNGE
9. *Accumulating Insecurity: Violence and Dispossession in the Making of Everyday Life*
   EDITED BY SHELLEY FELDMAN, CHARLES GEISLER, AND GAYATRI A. MENON
10. *They Saved the Crops: Labor, Landscape, and the Struggle over Industrial Farming in Bracero-Era California*
    BY DON MITCHELL
11. *Faith Based: Religious Neoliberalism and the Politics of Welfare in the United States*
    BY JASON HACKWORTH
12. *Fields and Streams: Stream Restoration, Neoliberalism, and the Future of Environmental Science*
    BY REBECCA LAVE
13. *Black, White, and Green: Farmers Markets, Race, and the Green Economy*
    BY ALISON HOPE ALKON
14. *Beyond Walls and Cages: Prisons, Borders, and Global Crisis*
    EDITED BY JENNA M. LOYD, MATT MITCHELSON, AND ANDREW BURRIDGE
15. *Silent Violence: Food, Famine, and Peasantry in Northern Nigeria*
    BY MICHAEL J. WATTS
16. *Development, Security, and Aid: Geopolitics and Geoeconomics at the U.S. Agency for International Development*
    BY JAMEY ESSEX
17. *Properties of Violence: Law and Land-Grant Struggle in Northern New Mexico*
    BY DAVID CORREIA
18. *Geographical Diversions: Tibetan Trade, Global Transactions*
    BY TINA HARRIS
19. *The Politics of the Encounter: Urban Theory and Protest under Planetary Urbanization*
    BY ANDY MERRIFIELD
20. *Rethinking the South African Crisis: Nationalism, Populism, Hegemony*
    BY GILLIAN HART
21. *The Empires' Edge: Militarization, Resistance, and Transcending Hegemony in the Pacific*
    BY SASHA DAVIS
22. *Pain, Pride, and Politics: Social Movement Activism and the Sri Lankan Tamil Diaspora in Canada*
    BY AMARNATH AMARASINGAM
23. *Selling the Serengeti: The Cultural Politics of Safari Tourism*
    BY BENJAMIN GARDNER
24. *Territories of Poverty: Rethinking North and South*
    EDITED BY ANANYA ROY AND EMMA SHAW CRANE
25. *Precarious Worlds: Contested Geographies of Social Reproduction*
    EDITED BY KATIE MEEHAN AND KENDRA STRAUSS
26. *Spaces of Danger: Culture and Power in the Everyday*
    EDITED BY HEATHER MERRILL AND LISA M. HOFFMAN
27. *Shadows of a Sunbelt City: The Environment, Racism, and the Knowledge Economy in Austin*
    BY ELIOT M. TRETTER
28. *Beyond the Kale: Urban Agriculture and Social Justice Activism in New York City*
    BY KRISTIN REYNOLDS AND NEVIN COHEN
29. *Calculating Property Relations: Chicago's Wartime Industrial Mobilization, 1940–1950*
    BY ROBERT LEWIS

30. *In the Public's Interest: Evictions, Citizenship, and Inequality in Contemporary Delhi*
BY GAUTAM BHAN

31. *The Carpetbaggers of Kabul and Other American-Afghan Entanglements: Intimate Development, Geopolitics, and the Currency of Gender and Grief*
BY JENNIFER L. FLURI AND RACHEL LEHR

32. *Masculinities and Markets: Raced and Gendered Urban Politics in Milwaukee*
BY BRENDA PARKER

33. *We Want Land to Live: Making Political Space for Food Sovereignty*
BY AMY TRAUGER

34. *The Long War: CENTCOM, Grand Strategy, and Global Security*
BY JOHN MORRISSEY

35. *Development Drowned and Reborn: The Blues and Bourbon Restorations in Post-Katrina New Orleans*
BY CLYDE WOODS
EDITED BY JORDAN T. CAMP AND LAURA PULIDO

36. *The Priority of Injustice: Locating Democracy in Critical Theory*
BY CLIVE BARNETT

37. *Spaces of Capital / Spaces of Resistance: Mexico and the Global Political Economy*
BY CHRIS HESKETH

38. *Revolting New York: How 400 Years of Riot, Rebellion, Uprising, and Revolution Shaped a City*
GENERAL EDITORS: NEIL SMITH AND DON MITCHELL
EDITORS: ERIN SIODMAK, JENJOY ROYBAL, MARNIE BRADY, AND BRENDAN O'MALLEY

39. *Relational Poverty Politics: Forms, Struggles, and Possibilities*
EDITED BY VICTORIA LAWSON AND SARAH ELWOOD

40. *Rights in Transit: Public Transportation and the Right to the City in California's East Bay*
BY KAFUI ABLODE ATTOH

41. *Open Borders: In Defense of Free Movement*
EDITED BY REECE JONES

42. *Subaltern Geographies*
EDITED BY TARIQ JAZEEL AND STEPHEN LEGG

43. *Detain and Deport: The Chaotic U.S. Immigration Enforcement Regime*
BY NANCY HIEMSTRA

44. *Global City Futures: Desire and Development in Singapore*
BY NATALIE OSWIN

45. *Public Los Angeles: A Private City's Activist Futures*
BY DON PARSON
EDITED BY ROGER KEIL AND JUDY BRANFMAN

46. *America's Johannesburg: Industrialization and Racial Transformation in Birmingham*
BY BOBBY M. WILSON

47. *Mean Streets: Homelessness, Public Space, and the Limits of Capital*
BY DON MITCHELL

48. *Islands and Oceans: Reimagining Sovereignty and Social Change*
BY SASHA DAVIS

49. *Social Reproduction and the City: Welfare Reform, Child Care, and Resistance in Neoliberal New York*
BY SIMON BLACK

50. *Freedom Is a Place: The Struggle for Sovereignty in Palestine*
BY RON J. SMITH

51. *Loisaida as Urban Laboratory: Puerto Rico Community Activism in New York*
BY TIMO SCHRADER

52. *Transecting Securityscapes: Dispatches from Cambodia, Iraq, and Mozambique*
BY TILL F. PAASCHE AND JAMES D. SIDAWAY

53. *Non-performing Loans, Non-performing People: Life and Struggle with Mortgage Debt in Spain*
BY MELISSA GARCÍA-LAMARCA

54. *Disturbing Development in the Jim Crow South*
BY MONA DOMOSH

55. *Famine in Cambodia: Geopolitics, Biopolitics, Necropolitics*
BY JAMES A. TYNER

56. *Well-Intentioned Whiteness: Green Urban Development and Black Resistance in Kansas City*
BY CHHAYA KOLAVALLI

57. *Urban Climate Justice: Theory, Praxis, Resistance*
EDITED BY JENNIFER L. RICE, JOSHUA LONG, AND ANTHONY LEVENDA

58. *Abolishing Poverty: Towards Pluriverse Futures and Politics*
BY VICTORIA LAWSON, SARAH ELWOOD, MICHELLE DAIGLE, YOLANDA GONZÁLEZ MENDOZA, ANA P. GUTIÉRREZ GARZA, JUAN HERRERA, ELLEN KOHL, JOVAN LEWIS, AARON MALLORY, PRISCILLA MCCUTCHEON, MARGARET MARIETTA RAMÍREZ, AND CHANDAN REDDY

59. *Outlaw Capital: Everyday Illegalities and the Making of Uneven Development*
BY JENNIFER LEE TUCKER

60. *High Stakes, High Hopes: Urban Theorizing in Partnership*
BY SOPHIE OLDFIELD

61. *The Coup and the Palm Trees: Agrarian Conflict and Political Power in Honduras*
BY ANDRÉS LEÓN ARAYA

62. *Cultivating Socialism: Venezuela, ALBA, and the Politics of Food Sovereignty*
BY ROWAN LUBBOCK

63. *Green City Rising: Contamination, Cleanup, and Collective Action*
BY ERIN GOODING

64. *New Destinations of Empire: Mobilities, Racial Geographies, and Citizenship in the Transpacific United States*
BY EMILY MITCHELL-EATON

65. *Spaces of Anticolonialism: Delhi's Anticolonial Governmentalities*
BY STEPHEN LEGG

66. *Migrant Justice in the Age of Removal: Rights, Law, and Resistance against Territory's Exclusions*
BY JACOB P. CHAMBERLAIN

67. *The Injustice of Property: Homeless Encampments and the Limits of Liberalism*
BY STEPHEN PRZYBYLINSKI

68. *Dispersed Dispossession: Collective Goods, Appropriation, and Agency in Rural Russia*
BY ALEXANDER VORBRUGG

69. *All Geographers Should Be Feminist Geographers: Creating Care-Full Academic Spaces*
BY LINDSAY NAYLOR WITH EMERALD L. CHRISTOPHER, EDEN KINKAID, CAROLINE FARIA, AND LATOYA E. EAVES

70. *Infrastructures of Caring Citizenship: Commoning Social Reproduction in Crisis-Ridden Athens, Greece*
BY ISABEL GUTIÉRREZ SÁNCHEZ

71. *Conservation in Common: Managing Wildlife and Sustaining Community on the Maasai Steppe*
BY JUSTIN RAYCRAFT

www.ingramcontent.com/pod-product-compliance
Lightning Source LLC
Chambersburg PA
CBHW020814230426
43666CB00007B/1002